Vintage Telephones of the World

P.J. Povey
&
R.A.J. Earl

Peter Peregrinus Ltd. in association with the Science Museum, London

Published by: Peter Peregrinus Ltd, London, United Kingdom

While the author and the publishers believe that the information and guidance given in this work is correct, all parties must rely upon their own skill and judgment when making use of it. Neither the author nor the publishers assume any liability to anyone for any loss or damage caused by any error or omission in the work, whether such error or omission is the result of negligence or any other cause. Any and all such liability is disclaimed.

ISBN 0 86341 140 1

Printed in England at The Alden Press, Osney Mead, Oxford

IEE History of Technology Series 8

Series Editor: Brian Bowers

Vintage Telephones of the World

Other volumes in this series

Volume 1 Measuring instruments—tools of knowledge
P. H. Sydenham

Volume 2 Early radio wave detectors
V. J. Phillips

Volume 3 A history of electric light and power
B. Bowers

Volume 4 The history of electric wires and cables
R. M. Black

Volume 5 An early history of electricity supply
J. D. Poulter

Volume 6 Technical history of the beginnings of radar
S. S. Swords

Volume 7 British television. The formative years
R. W. Burns

Acknowledgement

While preparing this book, we have been helped by many people in many lands throughout the world. So many, indeed, that a list of them all would be almost endless. We therefore make this collective acknowledgement, offering each one our grateful and sincere thanks.

R.A.J. Earl
P.J. Povey

Biographical notes on authors

Reg. Earl was the Curator of the British Telecom Museum in Oxford for 25 years. He is the author of 'The Development of the Telephone in Oxford', published by British Telecom, and now in its second edition. He is a Freeman of the City of Oxford and the Freemen's Archivist. Before becoming a curator he was a telephone engineer and served during the War as an R.A.F. Navigator. He lives in Oxford.

Peter Povey was for over 28 years the Curator of the British Telecom Museum in Taunton. In 1974 he was awarded the B.E.M. for his services to telephone history. He is the author of 'The Telephone and the Exchange', published by Pitman and also translated and published in Spanish. Before becoming a curator, he worked as a telephone engineer and served during the War in the Royal Electrical and Mechanical Engineers. He has now left British Telecom and lives on the Isle of Wight and works as a telecommunications consultant.

Contents

Introduction

This book was written with two objectives. Firstly to tell the story of the telephone instrument in an interesting and entertaining way. This was not difficult because the story contains many interesting episodes. The greatest problem was selecting the items to be included from the wealth of information unearthed. Almost all technicalities have been omitted. The few exceptions are those which have influenced telephone design and then, the explanation has been kept brief and simple.

The second objective was to tell the story of the telephone instrument from a truly international standpoint. Indeed, in the opinion of the authors, this is the only way in which the complete story can be told.

The idea of the electric telephone was first thought of by a Frenchman – Charles Bourseul. Later, a German, Philipp Reis, was the first person to transmit sound. The first telephone to reproduce intelligible speech was devised by Alexander Graham Bell, who was Scottish by birth but received his inspiration while living in Canada and, by the time he succeeded in making his idea work, was living in the USA. Since then, the ideas and achievements of men from practically every developed country have contributed to the telephone as we know it.

It is hoped that this book will be informative and that it will give an insight into the problems of telephone design but, most of all, it is hoped that people will enjoy reading it.

Chapter 1

The beginning of the telephone

Historically, the word 'telephone' has been used as a name for a number of devices ranging from a marine fog-horn to a multiple telegraph system. Today, a 'telephone' is an electrical device capable of transmitting and receiving speech. The first person to conceive such a device was a Frenchman, Charles Bourseul who, in 1854, wrote:

> . . . I wonder whether electricity could transmit articulate speech. In other words, if it would be possible to speak in Vienna and be heard in Paris. This is feasible in the following manner.
>
> Sounds, as is well known, are composed of vibrations which are reproduced by intermediary media. These vibrations carry sound to the ear. But the intensity of the vibrations weakens very quickly with distance, so that it is impossible to exceed quite narrow limits, even by means of megaphones, speaking tubes and ear-trumpets.
>
> Imagine you are speaking close to a moving plate so flexible as to register all voice vibrations. Now imagine this plate alternately makes and breaks the connection with a battery and that you have, at a distance, another plate producing simultaneously the same vibrations.
>
> Whatever happens, it is certain that in the near future electricity will transmit articulate speech at a distance.[1]

Bourseul made no attempt to construct equipment to do this and he did not use the word 'telephone'.

The first practical experimenter was Philipp Reis,[2] a teacher in Frankfurt-on-Main. He constructed several devices which transmitted distorted sound. He called his equipment a 'telephon' (*sic*) and demonstrated it publicly in 1860. In his transmitter a diaphragm, moved by sound, broke electrical contacts in a battery circuit. For his receiver he used a bar electromagnet attached to a sounding board. This utilised the 'click' produced by an electromagnet when it is magnetised – a phenomena which is called the 'Page effect'.

Over the years many notable scientists and historians have tried to evaluate the

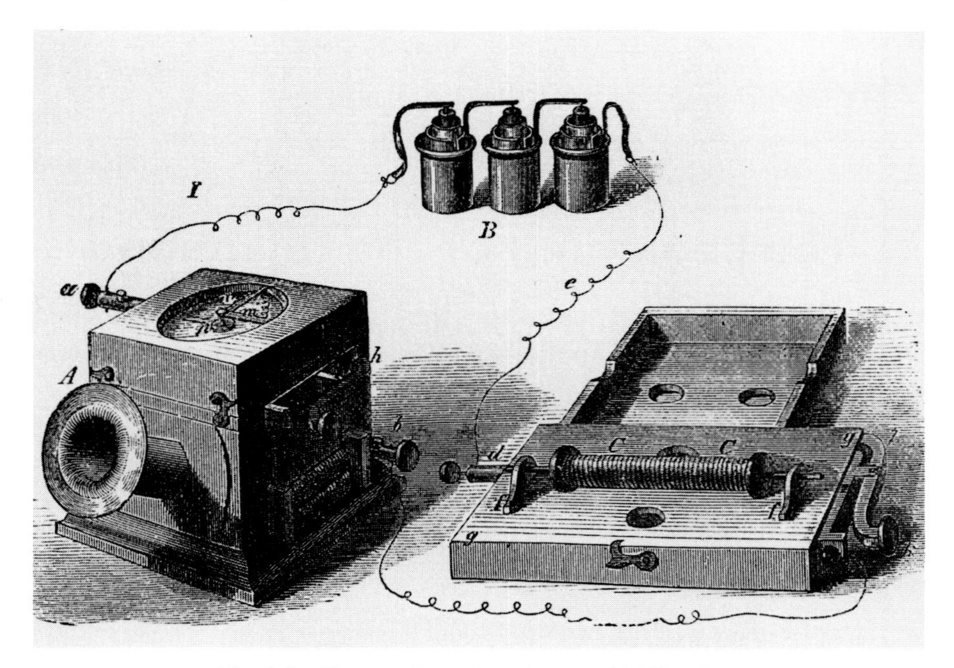

Fig. 1.1 *The experimental equipment of Philipp Reis*

results Reis obtained but, for a number of reasons, they arrived at conflicting conclusions. The results Reis might have achieved depended not only on the equipment he used but also in the way in which he used it. Unfortunately, historical records leave many important questions unanswered. Furthermore, while we know that the sound Reis transmitted was distorted, there is no consensus on the amount of distortion which would have been tolerable. In a contemporary description the Reis telephone was said to have sounded like a toy trumpet. It was looked upon as an interesting scientific experiment but no attempt was made to use it commercially. Today, most historians believe the Reis telephone was incapable of transmitting intelligible speech.

The first person to publicly demonstrate a telephone clearly capable of transmitting and receiving speech was Alexander Graham Bell. He was Scottish by birth but emigrated to Canada in his early twenties, and it was there, in 1874, that he conceived the idea of the telephone. Later he worked in Boston, Massachusetts, where he first succeeded in transmitting speech on 10 March 1876. In his early experiments he generated a varying electric current by electromagnetism and used a similar electromagnetic device as a receiver.

But for his experiment on 10 March 1876 he used a liquid transmitter with an electromagnetic receiver. The liquid transmitter was powered by a battery and utilised the variation in resistance which occurred when the depth of a needle dipping into a container of water and sulphuric acid was varied by the movement

Fig. 1.2 *Bell's second membrane telephone of 1875, often called the 'Gallows' telephone because it is shaped like an old fashioned gibbet*

of a membrane, or diaphragm, in response to sound. The only record of the appearance of this transmitter is a rough sketch in Bell's notebook.

Although the liquid transmitter was successful, Bell subsequently reverted to his earlier idea and it was the electromagnetic telephone with similar transmitter and receiver that he developed for commercial use.

Once Bell was satisfied that the telephone would work, he tried several designs. The most successful was the butterstamp telephone, so-called because it resembled the wooden stamps then commonly used for impressing designs on pats of butter.

For basic telephone service all that was needed was a line with a butterstamp telephone at each end. The instrument was held to the mouth or the ear depending on whether the person wished to speak or to listen. This method of use had the advantage that it was similar to the way in which the long-established speaking tube was used.

Even so, some people had difficulty in co-ordinating the movement of the

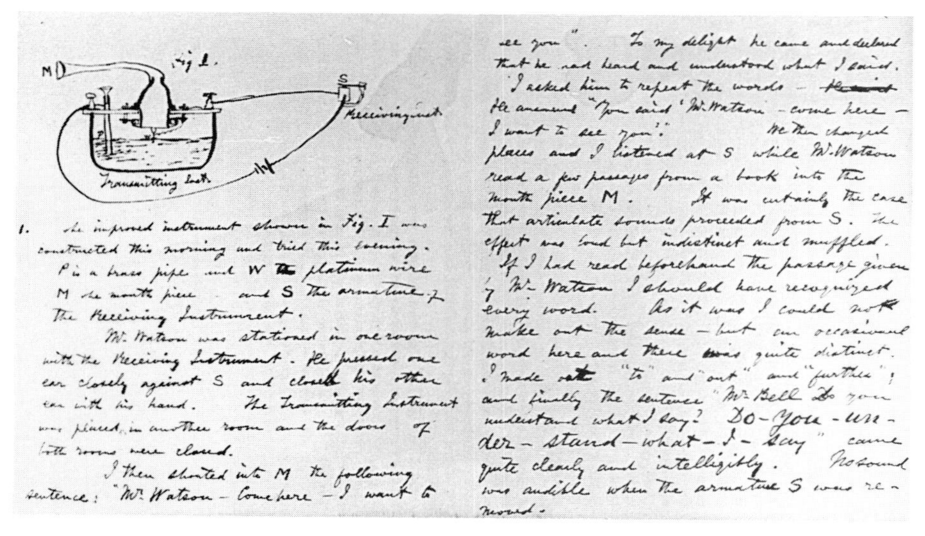

1. The improved instrument shown in Fig. I was constructed this morning and tried this evening. P is a brass pipe and W the platinum wire M the mouth piece and S the armature of the Receiving Instrument.

Mr. Watson was stationed in one room with the Receiving Instrument. He pressed one ear closely against S and closed his other ear with his hand. The Transmitting Instrument was placed in another room and the doors of both rooms were closed.

I then shouted into M the following sentence: "Mr. Watson – Come here – I want to see you". To my delight he came and declared that he had heard and understood what I said.

I asked him to repeat the words – He answered "You said 'Mr. Watson – come here – I want to see you.'" We then changed places and I listened at S while Mr. Watson read a few passages from a book into the mouth piece M. It was certainly the case that articulate sounds proceeded from S. The effect was loud but indistinct and muffled.

If I had read beforehand the passage given by Mr. Watson I should have recognized every word. As it was I could not make out the sense – but an occasional word here and there was quite distinct. I made out "to" and "out" and "further"; and finally the sentence "Mr. Bell do you understand what I say? Do-you-un-der-stand-what-I-say" came quite clearly and intelligibly. No sound was audible when the armature S was removed.

Fig. 1.3 *Bell's notebook entry for 10 March 1876 with sketch of his liquid transmitter*

telephone with their alternate speaking and listening. For these people the solution was to have two instruments so that one could be permanently held to the mouth and the other to the ear.

The earliest method of calling was to tap the diaphragm with the blunt end of a pencil. This produced a relatively loud sound at the distant end of the line. Watson, Bell's assistant, mechanised this process to some extent by making a telephone incorporating a tiny hammer which struck the diaphragm when a button was pressed. Subsequently he invented the telephone bell.[3]

The first butterstamp telephones were made in the workshop of Charles Williams Jnr. of Boston, Massachusetts, USA, and telephones imported from this manufacturer were the earliest to be used in Europe. But European manufacturers quickly started producing their own. In Britain, butterstamp telephones, closely resembling the original design, were made under licence by the Consolidated Telephone Construction and Maintenance Company and also by the India Rubber, Gutta Percha and Telegraph Works Company. In Germany, Werner von Siemens made telephones with longer and more efficient magnets which improved transmission, particularly on longer lines. In Sweden, Lars Magnus Ericsson manufactured instruments in which a conventional Bell butterstamp telephone was used as a receiver while the Siemens modified version was used as a transmitter.

Various methods of calling were also tried. Frederic A. Gower, an American working in France, devised a calling device consisting of a harmonium reed mounted on the back of the telephone diaphragm. To operate the device the caller blew into the instrument. Air passed through a hole in the diaphragm

Fig. 1.4 *The base of one of the bronze lamp standards modelled by the Victorian sculptor, William S. Frith, which flank the entrance of number 2 Temple Place, London. The design symbolises the marvels of telephony, telegraphy and electrical illumination. Telephony is represented by two cherubs conversing via butterstamp telephones*

Fig. 1.5 *This telephone, with wall mounted bell and press button for calling, had hooks for two butterstamp instruments. It was one of the first to be used with telephone exchanges*

vibrating the reed and, with it, the diaphragm itself, thereby causing a sound of considerable volume to be reproduced by the distant telephone. This, and some other contemporary devices, had the advantage that, like the butterstamp telephone itself, they were operated in a similar way to the speaking tube, with which people were already familiar. Gower sold telephones in several European countries, most notably France and Portugal.

In Sweden, L.M. Ericsson made telephones with a small detachable signal trumpet. To make a call the trumpet was fitted into the telephone and blown. It was then removed so that the telephone could be used for speaking.

Telephones with an assortment of whistles and rattles were also made by Siemens in Germany. One, in particular, had a detachable mouthpiece which

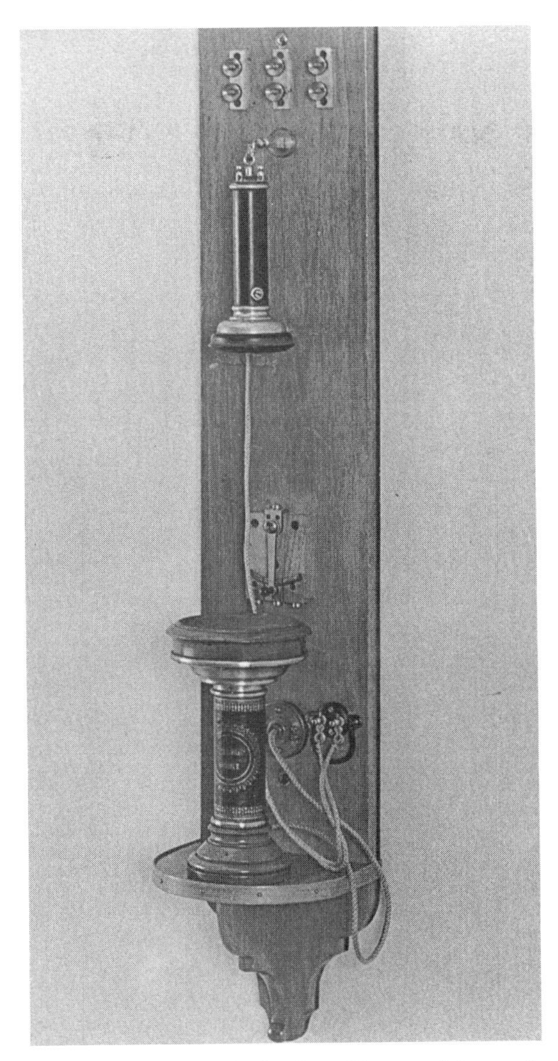

Fig. 1.6 *The first wall telephone made by Lars Magnus Ericsson in 1880, in which a conventional Bell butterstamp telephone was used as a receiver and the larger Siemens modified version of the Bell telephone was used as a transmitter*

contained both a reed and a metal ball. When the mouthpiece was fitted into the telephone and blown, the reed produced a sound while the ball was made to dance up and down on the diaphragm.

In the USA, Watson's invention of the telephone bell was adopted so rapidly that horns, whistles and rattles, as a means of calling, were very little used, and this was also true in Britain where American practice was closely followed.

Fig. 1.7 *Desk telephone with removable signal trumpet made by Lars Magnus Ericsson in 1879*

Fig. 1.8 *Telephone with rattle calling (left) and whistle calling (right) made by Siemens in 1878*

Fig. 1.9 *Early evolution in telephone design is well illustrated by this Siemens and Halske instrument of 1885. Two identical butterstamp telephones are still used but the one intended for speaking has been partially boxed in to produce a wall telephone*

The telephone bell soon became the most widely used calling device, although it needed an electrical circuit. To avoid having a separate line wire, Watson used the same line for both ringing and speaking. This was possible because there would be no one speaking when the bell was needed.

To make the single line do both jobs, Watson used a manual change-over switch. This was shown in his patent specification[3] and used in the earliest telephones with bell-calling. In the hands of the public, however, these switches proved to be a source of trouble. People forgot to restore them to the 'bell' position after a call. This left the bell disconnected and prevented further calls being received.

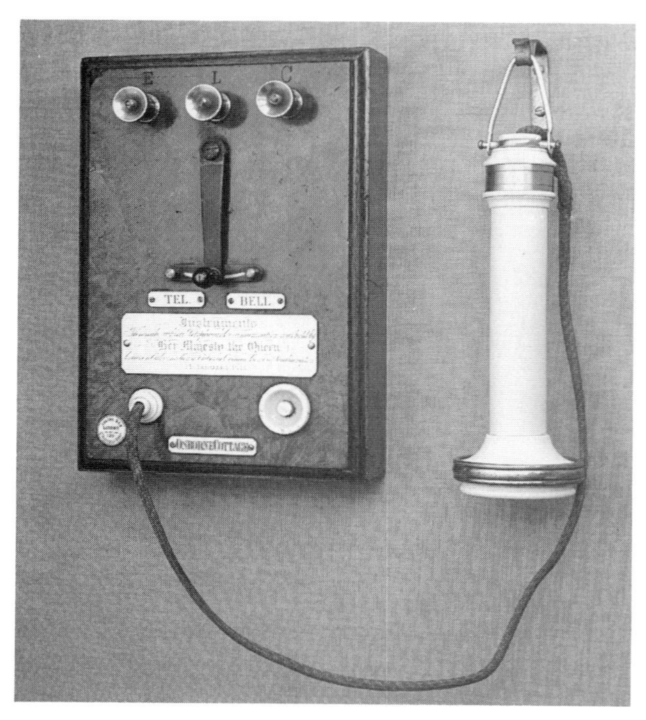

Fig. 1.10 *Bell telephone of 1878 with change-over switch made by Julius Sax and Company, of London, and presented by Alexander Graham Bell to Queen Victoria. This telephone was installed at the Osborne Cottage end of a private line to her nearby residence, Osborne House*

To overcome this problem, an American, H.L. Roosevelt, invented the 'gravity switch'. This used the weight of the receiver to operate the change-over switch. A typical way of doing this was to use the movement of a spring-loaded hook on which the receiver was hung. Roosevelt patented his invention and, to avoid paying royalties, G.M. Phelps, also of America, devised an alternative device. Phelps' alternative consisted of a clasp being used to operate the change-over switch.[4]

When he first transmitted speech in 1876, Bell used a variable resistance liquid transmitter. Although he later abandoned this idea in favour of the electromagnetic transmitter, it was with the variable resistance principle that future progress lay. This was because the improvement which could be achieved with the electromagnetic transmitter was limited. It was a power converter and the output could never exceed the modest amount of power in the sound input. The variable resistance transmitter was not restricted in this way because it was supplied with power from a battery.

The first step towards a practical variable resistance transmitter was taken in

Fig. 1.11 *Close up of mechanism of Phelps receiver holder, or clasp, on telephone installed in New Zealand*

1877 when Emile Berliner, an emigrant from Hanover, Prussia, working in the USA, devised the first non-liquid transmitter of this type. It consisted of an iron diaphragm touching a steel ball.[5] Berliner's invention was never used commercially because it was quickly followed by a better variable resistance transmitter designed by the famous American inventor, Thomas Alva Edison. In his resistance element Edison used carbon, a material with properties which make it without equal for this purpose.

The way in which the telephone developed in the period following the invention of the carbon transmitter was governed as much by legal as by technical considerations. In the USA, telephones comprising a carbon transmitter and an electromagnetic receiver were used at a very early date. But in Britain there were different patents, with different wording, and different interpretations.

Fig. 1.12 *Edison's telephone with carbon transmitter and chalk receiver, 1879*

Edison was advised by his London agent that the British patent for his transmitter was worthless without a receiver free from Bell patents. Edison succeeded in inventing a new receiver in a remarkably short time and patented it in 1879.[6]

Edison's receiver exploited one of his own discoveries. Some years earlier he had noticed that in certain circumstances the friction of one body moving against another varied when an electrical current was passed between them. The receiver he designed contained a drum of chalk moistened with potassium iodide. Upon this drum was pressed a platinum stud joined by a rod to the centre of a diaphragm. Throughout the telephone call the drum had to be rotated and a handle was provided for this purpose. The drag between the revolving drum and the stud tensioned the diaphragm, and this tension was varied by the electrical input which was connected between the drum and the stud. The diaphragm responded by moving backwards and forwards reproducing the original sound.

This receiver was known as a 'motograph', an 'electromotograph', or simply as a 'chalk receiver'. It did not convert power because the diaphragm was driven by the turning handle and it is said to have been capable of producing enough volume to fill a small hall.

The chalk receiver was first used commercially in Britain, in the late summer of 1879, where rival Bell and Edison companies had been set up. Edison claimed the

Fig. 1.13 *The Blake transmitter was sometimes incorporated in a complete telephone and sometimes mounted in a metal body but in the design most frequently used it was fitted in a small wooden case. These cases were of a standard pattern which remained unchanged for many years. They are easily recognised even without the words 'Blake transmitter' stamped on the front. However, the contents do not always match the label. Cases are often found containing replacement transmitters of more recent date*

volume of his telephone as a virtue but telephone users found it excessive and looked upon the need to constantly turn the handle throughout the call as a nuisance.

An interesting contemporary account of the Edison telephone is to be found in the work of the author, Mr. George Bernard Shaw, who was employed by The Edison Telephone Company of London Ltd. as a wayleave manager:

> You must not suppose, because I am a man of letters, that I never tried to earn an honest living. I began trying to commit that sin against my nature when I was fifteen and persevered from youthful timidity and diffidence until I was twenty-three. My last attempt was in 1879, when a company was formed in London to exploit an ingenious invention by Mr. Thomas Alva Edison – a much too ingenious invention as it proved being nothing less

Fig. 1.14 *Telephone of 1879 with Blake transmitter and Bell receiver made by the India Rubber Gutta Percha and Telegraph Works Company of London, and known as a 'Silvertown Set' from the name of the factory where it was made*

> than a telephone of such stentorian efficiency that it bellowed your most private communications all over the house instead of whispering them with some sort of discretion. This was not what the British stockbroker wanted. . . .[7]

Edison produced his carbon transmitter by practical experiment. The first person to provide a full explanation and proof of the way in which it worked was David

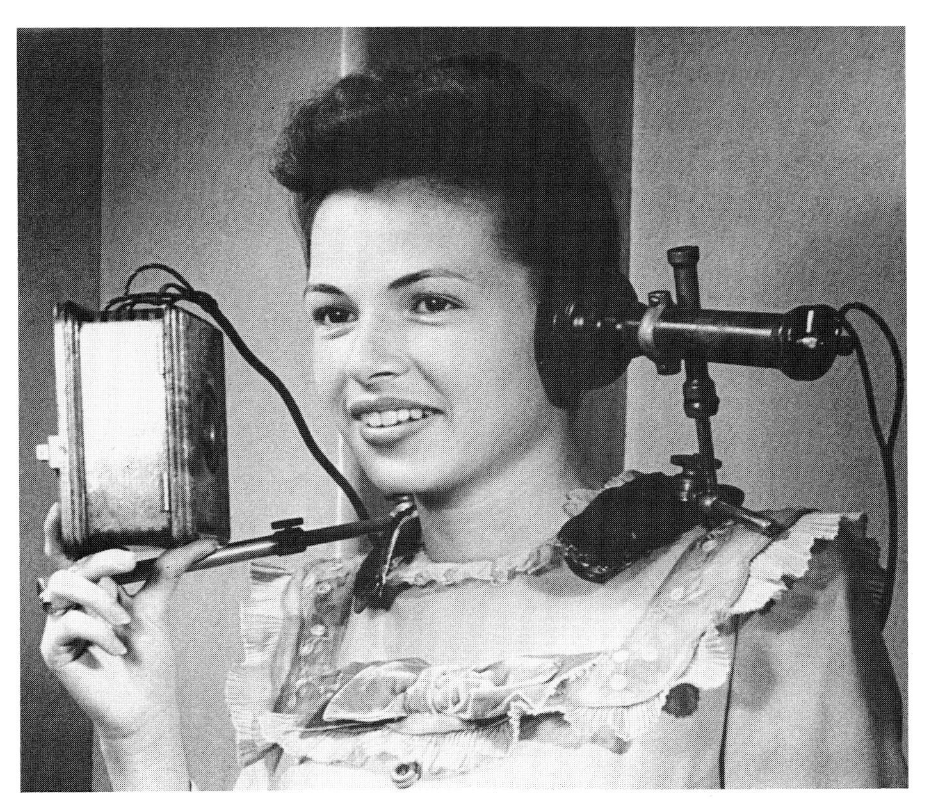

Fig. 1.15 *Perhaps the most unusual application of the Blake transmitter was the 'Gilliland Harness' invented by E.T. Gilliland in 1881. Its purpose was to enable telephone operators to work with both hands and it was the first operator's 'headset'. As it weighed over six pounds, it had to be supported on the shoulders*

Hughes. Born in London, of Welsh parents, he was taken to the USA at the age of seven. Later, he lived in France for some years before returning to London and it was there that he conducted his investigation into the operation of the carbon transmitter. He made no attempt to profit by patenting his ideas, instead he made his knowledge freely available by publishing an account of his experiments and the conclusions he had reached. This proved to be an enormous stimulus to transmitter design.[8]

Hughes demonstrated that the variations in resistance of a carbon transmitter, in response to sound, are not caused by compression of the carbon itself, as many people had believed, but are due to changes at the junctions between the component parts of the transmitter. He called this the 'microphonic effect'.

Although he did not produce a commercial transmitter himself, at least two of his experiments pointed the way which other inventors followed. In one

Fig. 1.16 *A telephone of 1880 sometimes called 'The Marriage' because it was one of the first telephones combining a carbon transmitter with an electromagnetic receiver. It was also one of the first pedestal telephones and a forerunner of the popular 'candlestick' telephone of later years*

experiment he used a carbon pencil sharpened to a point at each end and supported with its points resting in two carbon cups. This device functioned well as a transmitter. In another experiment he used substances broken into tiny particles. Both these experiments proved to be the beginning of important lines of development in telephone history.

The theoretical work of Hughes became the basis of a very successful transmitter designed by Francis Blake of the USA. His patent was acquired by the Bell company and the transmitter was widely used in America. It was also widely used in Britain although its legal status was in doubt because it used carbon. It was therefore seen as a version of the Edison transmitter although this was never tested in court because in 1880 the Bell and Edison interests in Britain merged.

The merger of the Bell and Edison interests made it possible to construct telephones from the best available components. Bell's electromagnetic telephone, when used as a transmitter, had a low output and was inferior to the carbon transmitter but, when used as a receiver, it worked efficiently and there was no need to turn a handle. The combination of carbon transmitter and electromagnetic receiver therefore became standard.

Chapter 2

Improved speaking and listening

One of the first of the many transmitters inspired by the work of David Hughes was designed by Louis John Crossley, a carpet manufacturer, of Halifax, England. Crossley's transmitter was based on Hughes' experimental carbon-pencil transmitter but, to increase its efficiency, Crossley used four loose carbon pencils. They were mounted in diamond formation between hollows in four carbon blocks secured to the underside of a diaphragm of very thin pine.[9]

Telephones incorporating Crossley transmitters were manufactured by Messrs. Blakey and Emmott of Halifax. This company's instruments were the first to be supplied to the British Post Office for its telephone service which commenced in certain places in 1881. These telephones were housed in a wall-mounted box with the diaphragm of the transmitter beneath a fretwork lid. The exchange was called by operating a press button on the right hand side of the box. This connected a battery to the line and, in keeping with the current telegraph practice, a galvanometer was fitted to indicate line current. A spring-loaded hook on the front of the box supported the characteristically dumpy Crossley receiver. The movement of this hook operated electrical contacts which made the circuit changes necessary at the beginning and end of calls.

The hook was a feature of many Victorian telephones and the origin of some of the expressions we still use today. For example, 'The telephone was off the hook', or 'I am going to hang up now'.

It was appreciated at a very early date that the efficiency of the Bell telephone receiver depended on the strength of the magnet. As early as June 1876 Bell had used horseshoe magnets which produce a more intense magnetic field. A few months later Watson read about laminated or compound magnets, which are composed of a number of thinner magnets clamped together, and he became aware that they could be strongly magnetised. He made telephones for Bell using laminated magnets and Bell showed a horseshoe magnet of this type in his second telephone patent of 1877.[10] This was, incidentally, the first patent in which Bell showed any sort of permanent magnet.

Although the need for a strong magnetic field was appreciated, the first butterstamp telephones to be used commercially had the less efficient straight,

Fig. 2.1 *Wall telephone with Crossley receiver made by Blakey and Emmott of Halifax in 1881. This was the first telephone made available to the public by the British Post Office*

unlaminated magnet because it was simpler, and cheaper, to manufacture. This was called the 'single pole' receiver. In 1880, improved single pole receivers with laminated magnets were introduced but it was not until the 1890s that Bell receivers with horseshoe magnets appeared. These were known as 'double pole' receivers. Like the earlier Bell receivers, they were designed to have a small diameter so that they could be conveniently grasped. This meant that the horseshoe magnet had to be narrowed into a hairpin-like shape.

In Germany, Siemens' version of Bell's butterstamp telephone used a horseshoe magnet from a very early date. Elsewhere in Europe, some telephone designers adopted the opposite approach. Instead of narrowing the horseshoe magnet into a hairpin shape, Ader, d'Arsonval, Goloubitzky, Ochorowicz and others broadened the magnet into an almost complete circle. This enabled it to be used both as a handle and as a means of hanging the receiver on the hook. The resultant shape must have been regarded as satisfactory because in several other designs, in which the magnet was enclosed within the body of the receiver, a ring was added which had no magnetic function. Instruments of this design are

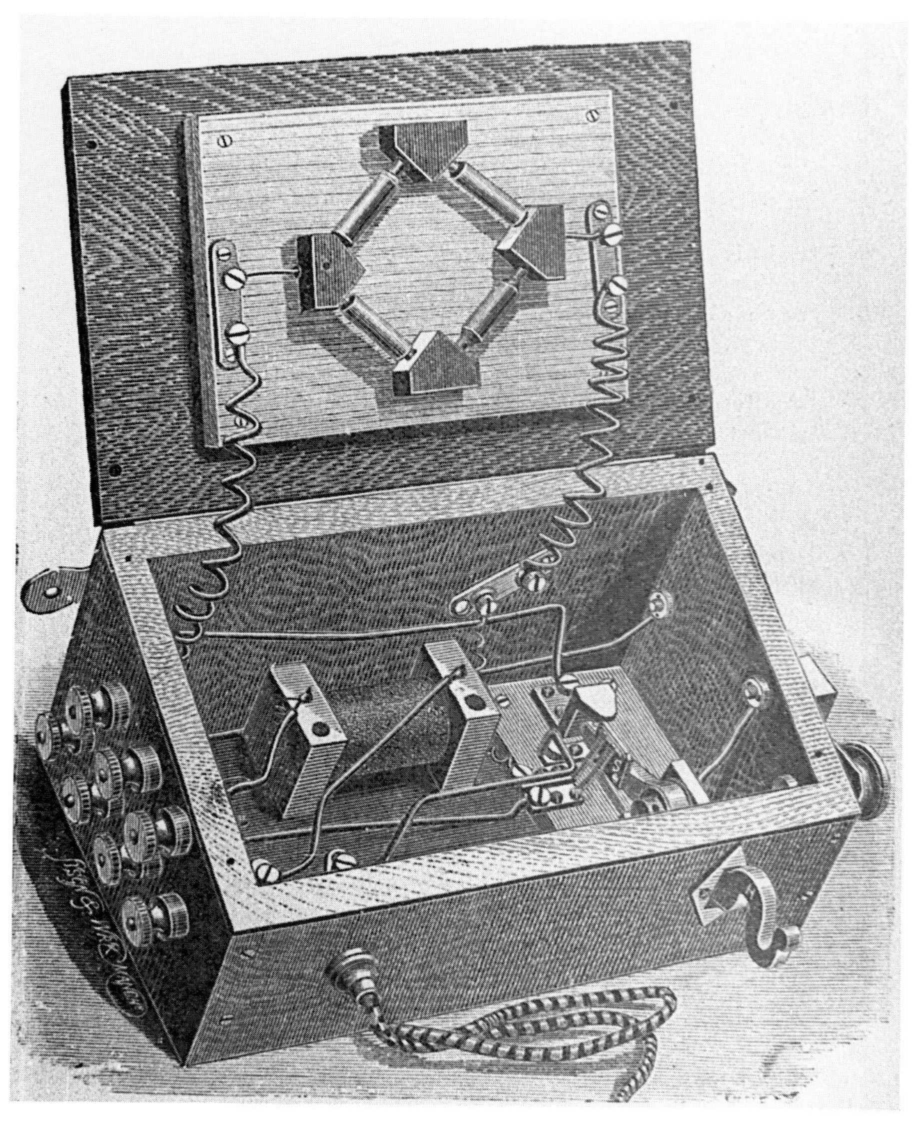

Fig. 2.2 *The Blakey and Emmott telephone of 1881 with lid open to show Crossley transmitter and switch-hook mechanism*

Fig. 2.3 *An Ader 'Bull Ring' receiver. The circular magnet was used both as a handle and as a means of hanging the receiver on the hook*

sometimes called 'bull ring' receivers irrespective of whether the ring functions as a magnet or not.

A third, totally different approach was used by Gower. He made no attempt to reduce either the size or the weight of his receiver. It was too big, and too heavy, to be conveniently held to the ear so he mounted it inside the case of the telephone and conveyed the sound to the user's ear by flexible tubes. In Gower's earliest telephones his receiver was also used as an electromagnetic transmitter but, from 1880 onwards, he used various carbon transmitters. The Gower telephone was the first to be adopted by the British Post Office on a large scale and followed its earliest telephone, the Blakey and Emmott. It was also used by other telephone administrations and several railway companies.

Crossley's carbon pencil transmitter was quickly followed by similar types in which various numbers of pencils in many different arrangements were used. In Britain, a transmitter with only two pencils was designed by William Johnson, one of the few early telephone inventors who was a professional telephone man. He was originally employed by the Telegraph Department of the British Post Office in the Mansfield district. Later, he became the manager of the Sheffield Telephone Exchange and Electric Light Company.

Johnson transmitters were exceptional in that they used only two pencils. Most inventors believed they could increase the efficiency of the carbon pencil transmitter by adding more pencils, thereby producing more junctions at which

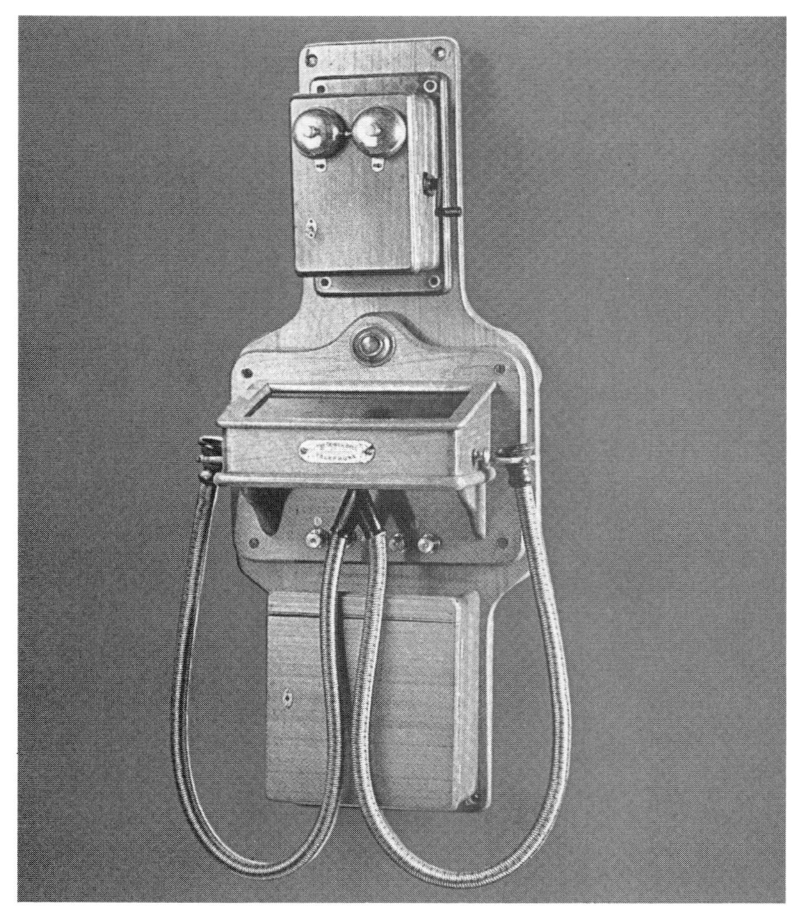

Fig. 2.4 *Gower telephone with carbon pencil transmitter – one of the earliest telephones to be used in Portugal*

current variations could occur. This was indeed the case, provided the sound caused similar variations at all the junctions and that these variations acted together to reinforce the output but, with this type of transmitter, this was not always achieved. Nevertheless, in mainland Europe, and particularly in France, innumerable arrangements of carbon pencils were tried. Probably the best known and the most successful was Ader's transmitter with ten pencils manufactured by the Société des Téléphones.

An English clergyman, the Reverend Henry Hunnings, of Bolton Percy in the county of Yorkshire, was responsible for the next development in the evolution of the telephone. He was not aware of the work of Hughes[11] but, even so, he invented and patented a transmitter in 1878[12] which was the logical development of Hughes' experiments with particles.

Fig. 2.5 *Telephone with circular Johnson transmitter of 1881*

A typical Hunnings' transmitter consisted of a platinum diaphragm with a brass plate slightly behind it, the intervening space being filled with powdered coke. Later, as the transmitter was developed, slightly larger particles, generally referred to as 'granules', were used. Hunnings' transmitter, and its successors, therefore became known as 'carbon granule' transmitters. Granules can be thought of as tiny carbon pencils which, because they are so small, can be concentrated in large numbers in a tiny space.

Seen in retrospect, a logical line of evolution can be traced from Edison's transmitter, via the transmitters of Blake and Crossley, to the carbon granule

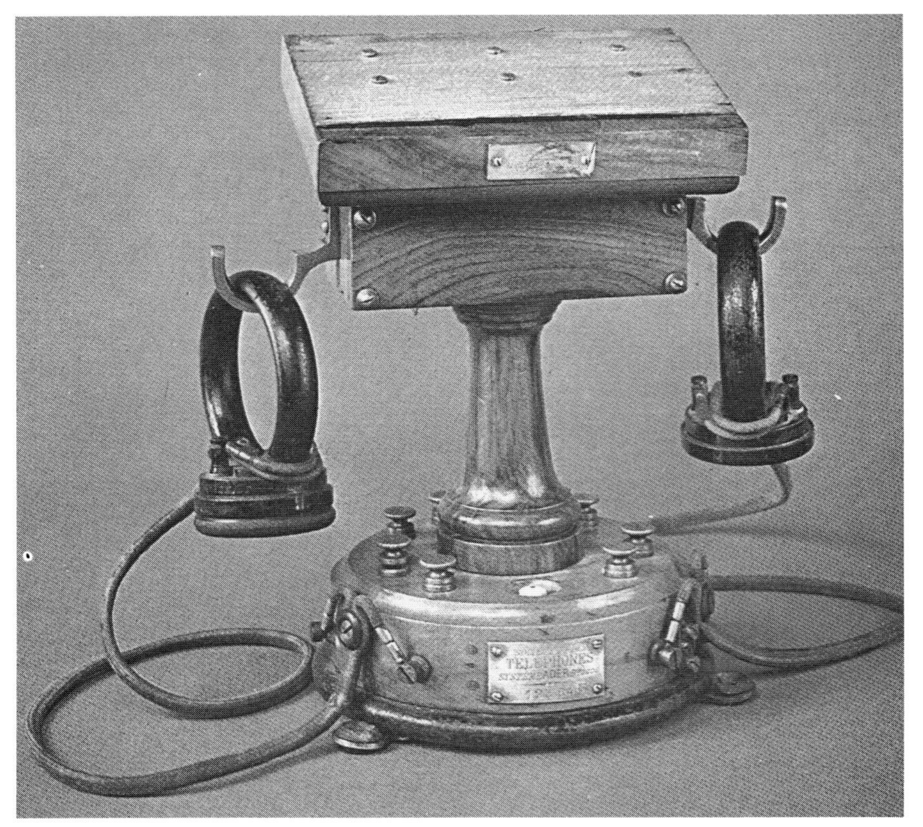

Fig. 2.6 *Ader table telephone. The caller spoke against the thin wooden top of the instrument beneath which carbon pencils were mounted*

transmitter of Hunnings, and it might be expected that each new transmitter would have displaced its predecessor, but this is not what happened. All the transmitters were capable of further improvement and it was by no means apparent which would ultimately emerge as the best. Hunnings' transmitter, in particular, suffered from a serious defect. Granules became packed in its lower part so solidly that it was difficult for the diaphragm to exert further pressure on them. Ultimately, this problem was overcome but, until then, the older types of transmitters not only continued to be used, they continued to be developed and improved.

In 1880, Lars Magnus Ericsson, of Sweden, designed a new version of the Blake transmitter. Like the original, it needed to be carefully and precisely adjusted. Blake had provided for this by incorporating in his design an adjuster consisting of a lever positioned by the pressure of a screw against a sloping face.

Fig. 2.7 *25 franc stamp of the Niger Republic showing Ader telephone with carbon pencil transmitter. Issued to celebrate the centenary of the International Telecommunications Union in 1965*

The new version was more compact and adjustments were made by means of a screw, or helix, and it was from this that it derived its name – the 'helical' transmitter.[13]

It was extremely compact, with the diaphragm and variable resistance element contained in a small capsule. A horn-shaped mouthpiece was mounted against the diaphragm and, opposite it, a brass adjusting screw protruded. The whole assembly was self contained and neat enough to be mounted outside the telephone.

Although it was not a major technical advance, the helical transmitter had a profound effect on the shape and general appearance of telephone transmitters and its influence can be seen in many later developments. This proved to be only one of a number of innovations which, over the years, originated in Scandinavia and, in all, the impact of Scandinavian ideas on telephone design has been out of all proportion to the size of the population. Nowhere – in Europe at least – are

Fig. 2.8 *Ader telephone of 1888 with switchboard for two lines. Fitted at Blenheim Palace, home of the Dukes of Marlborough and birthplace of Sir Winston Churchill*

people more telephone minded. This was noted as early as 1895 by an English traveller who wrote:

> Whenever two or three Swedes or Norwegians or Danes or Finns of Scandinavian descent are gathered together they almost infallibly proceed to immediately establish a church, a school and a telephone exchange.

In the USA, a telephone was introduced in the early 1880s which, during the

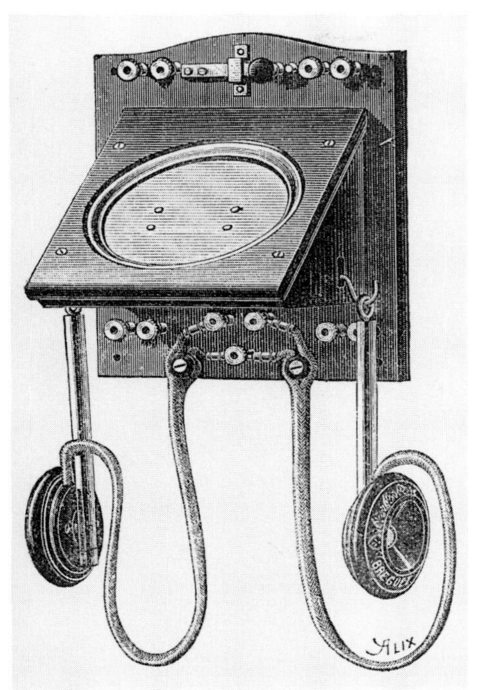

Fig. 2.9

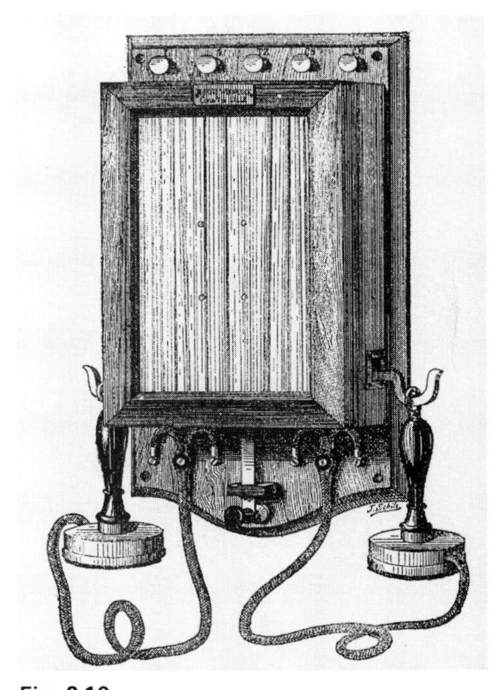

Fig. 2.10

Some nineteenth century telephones with carbon pencil transmitters

Fig. 2.11

Fig. 2.12

Fig. 2.13

Fig. 2.14

Some nineteenth century telephones with carbon pencil transmitters

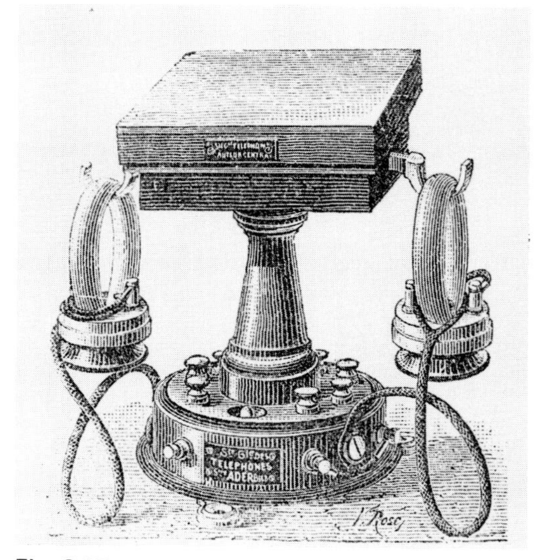

Fig. 2.15

Fig. 2.16 *The telephone Lars Magnus Ericsson designed in 1880 with helical transmitter and Bell receiver. Behind the transmitter can be seen the lightning protector consisting of two brass plates separated by a tiny gap across which sparks could easily jump to dissipate an electrical charge on the line*

next two decades, was destined to become extensively used in North America and in Britain. Indeed, it was used in every country which imported American telephones or where American design influenced local production. It was called the 'Blake' telephone although today it is almost invariably called the 'three box type'. It consisted of a sturdy backboard to which were fixed three boxes, one above the other. The lowest case contained the battery, and the lid of the case formed a small writing desk. The middle case contained a Blake transmitter while the upper case contained the bell, the magneto generator and the switch-hook. The magneto generator, or 'magneto' as it was frequently called, was used to

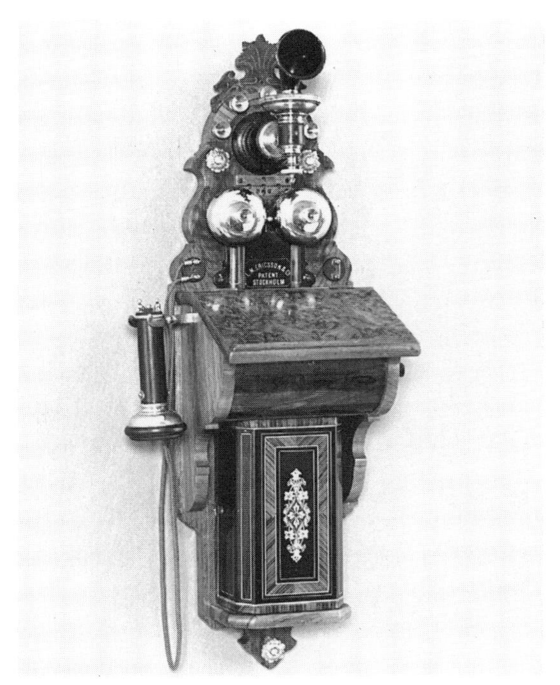

Fig. 2.17 *In 1882, L.M. Ericsson of Sweden brought together the latest in technology with the traditional craftsmanship to produce this magnificent work of art*

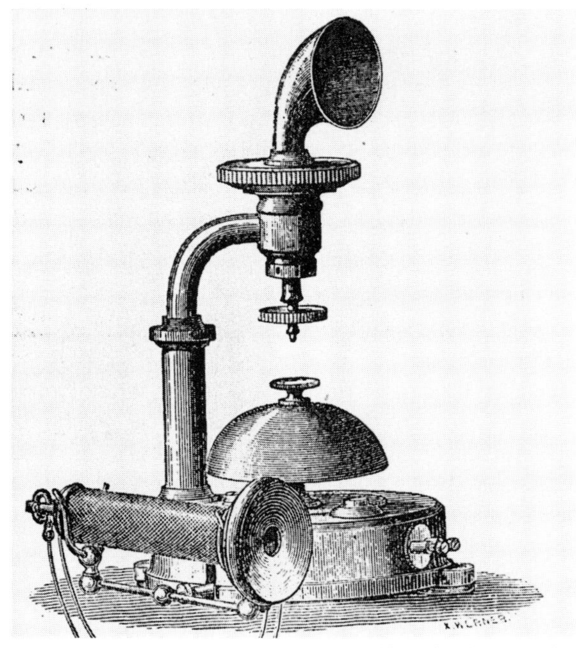

Fig. 2.18 *Telephone manufactured by L.M. Ericsson of Sweden in the early 1880s with helical transmitter and Bell receiver*

Fig. 2.19 *The three box telephone was introduced in the early 1880s and remained in use until the end of the century*

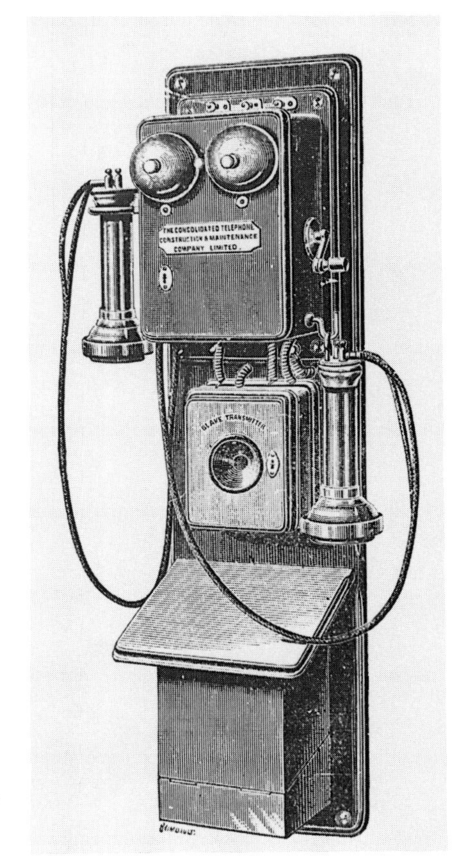

Fig. 2.20 *A three box telephone of the 1890s with twin Bell receivers manufactured by the Consolidated Telephone Construction and Maintenance Company Ltd*

Fig. 2.21 *This telephone was designed by L.M. Ericsson of Sweden in the mid 1880s. Bell's invention was still relatively new and there were few established conventions governing a telephone's appearance. Designers could draw inspiration from where they pleased. Ericsson was apprenticed to a blacksmith and for a time worked for the railway. Could it be that this remarkable instrument was inspired by the boiler, fly-wheel and safety valve of a steam engine?*

Fig. 2.22 *This version of the Ericsson table telephone of the 1880s has two receivers for people who prefer to listen with both ears or to enable a third person to join in the conversation*

Fig. 2.23 *Transmitter with horizontal diaphragm, patented in 1886 by Emile Berliner and used in the USA to improve long distance telephone calls*

generate an alternating current to ring the bell of the distant telephone or to operate a calling indicator at the exchange. This type of alternating current generator was adapted to the needs of the telephone by Thomas Watson – Bell's assistant.

The three box telephone was deservedly successful because it was a well thought-out, effective design. Separating the battery from all the other equipment avoided any corrosion the battery might cause. In the same way, because the Blake transmitter was dependent on careful adjustment and was easily upset by vibration, it was better housed in a separate box away from the bell and magneto. Later, the Blake transmitter was superseded by other, less delicate types but, by then, the three box telephone was so firmly established that it continued to be used in various forms until the end of the 19th century.

Fig. 2.24 *To prevent packing of the carbon granules, the transmitter of this telephone was rotated by a ratchet and pawl mechanism operated by the movement of the switch-hook*

Fig. 2.25 *Telephone with Berthon transmitter and Ader receiver installed in Blenheim Palace, England about 1888*

Fig. 2.26 *Table telephone with granular transmitter manufactured by Milde of Paris in 1892*

During the twenty years following the invention of the granular transmitter, many inventors attempted to perfect it and, in particular, to overcome the problem of the packing of the granules. In 1886, Emile Berliner, working on behalf of the Bell Company in America, patented a transmitter with a horizontal diaphragm. This was an attempt to avoid the uneven distribution of the granules due to gravity. They rested on the upper surface of the diaphragm and the plate which had been at the back of the granules was now on top. The sound was conveyed to the underside of the diaphragm by a horn.[14]

A different approach to the problem was to redistribute the granules by mechanical means. Telephones were made with the transmitter mounted on a wedge-shaped piece of metal which fitted into a 'V' shaped socket on the telephone. The transmitter was connected to the main body of the telephone by flexible wires. This arrangement enabled the user to remove the transmitter from the telephone and shake it whenever he felt it would be beneficial.

A variant of this idea, which was widely used, was to mount the transmitter in a circular capsule which could be rotated. Connections were made via slipping contacts or by flexible wires. In the latter case, a stop was fitted to prevent the capsule being repeatedly turned in the same direction.

Several inventors attempted to agitate the granules automatically by, for

Fig. 2.27 *The two receiver version of the British Post Office 'Universal' telephone of 1889 with Deckert transmitter. This telephone was called 'Universal' because the internal wiring could be adapted to any of the circuits then in use*

example, causing the switch-hook to jolt the transmitter when the receiver was removed. In one ingenious device the movement of the switch-hook was used to rotate the transmitter by a ratchet and pawl mechanism.

Other inventors sought to overcome the problem of packing by redesigning the transmitter itself. Countless experiments were made and many modifications were patented. Some of these ideas were used commercially and gave good results in their day. In Britain, Charles Moseley, of Manchester, patented a transmitter[15] that was used by the Post Office. In mainland Europe, granular transmitters were designed by Berthon, Ericsson, Milde, Roulez and others. In the USA, as well

Fig. 2.28

Fig. 2.29

Early telephones with carbon granular transmitters

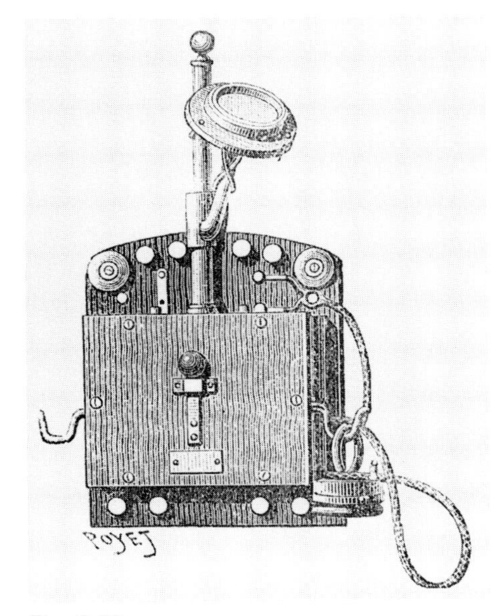

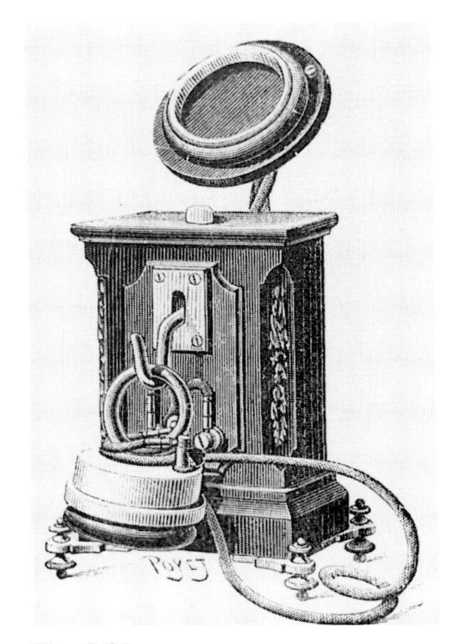

Fig. 2.30

Fig. 2.31

Fig. 2.32 **Fig. 2.33**

Early telephones with carbon granular transmitters

Fig. 2.34

Fig. 2.35 *This telephone, which would not be out of place in a palace, was manufactured by L.M. Ericsson of Sweden in 1882*

as the transmitter designed by Berliner, which has already been described, a transmitter was designed by Thornberry which also had a horizontal diaphragm. This transmitter was intended for use on long lines.

By far the most successful of the early granular transmitters was the type invented by W. Deckert of Austria.[16] Deckert's transmitter was manufactured by the General Electric Company and used by the British Post Office to replace the Gower transmitter. It was also used by the private telephone companies in Britain to replace the Blake transmitter. It was quite often fitted inside the box of the Blake transmitter so that today, it can never be assumed that a Blake box contains a Blake transmitter.

Packing was never absolutely eliminated but was eventually reduced to insignificance, not as the result of any major innovation, but by painstaking research into details of design such as the best size for granules, and the best size and shape for the chamber.

Chapter 3

Many shapes and sizes

It was the granular transmitter which made the telephone handset practical although the basic idea was much older and originated when the telephone was in its infancy.

As early as 1877, two Englishmen, Charles E. McEvoy and G.E. Pritchett, acting independently, patented handsets. McEvoy's patent describes two types. The first was basically a butterstamp telephone with a speaking tube attached. With the telephone held to the ear, the tube was of such a length and shape that the open end could be positioned in front of the mouth. For his second handset McEvoy suggested using two butterstamp telephones joined by a curved handle. In Pritchett's specification there is a drawing which looks remarkably like a modern handset. However, there is nothing to indicate how such a handset could have been made with the components then available and the idea is mainly of interest as a remarkable prediction of things to come.

The first handset to be actually made was designed in the following year, 1878, by R.G. Brown of the USA.[17] Brown provided for the bulky magnet, which was then indispensable to the Bell receiver, in a very ingenious way. He attached it to the back of the earpiece, at right-angles, and used it to form the handle joining the earpiece to the mouthpiece. He was also able to use the newly-invented Edison carbon transmitter although its susceptibility to vibration must have been a drawback. A few Brown handsets were used at the Gold and Stock Exchange in New York but the Bell company did not approve of handsets for use with its telephones and almost half a century was to pass before they were adopted for general use in the USA.

In search of a place where his ideas would be more readily accepted, Brown travelled to France and became the Electrical Engineer of the Société Générale des Téléphones. This organisation employed two other telephone pioneers, Berthon who was manager, and Ader. Both men devised telephone transmitters and receivers which bear their names. The Société Générale des Téléphones produced its first handset in 1879 and other European manufacturers quickly

Fig. 3.1 *The first Swedish handset, made by Anton Avén and Leonard Lundqvist in 1884, was crudely constructed from a Bell receiver, a helical transmitter and a rough wooden handle held together with string and wire. This model, in the historical collection of Sweden's Royal Telegraph Department, is in marked contrast to the beautifully constructed models sometimes seen in museums which are reputed to be the first of their kind. However, it is easy to believe that this handset is typical of the first equipment made by pioneers and that the well constructed models came later. The first commercial Swedish handsets were made in L.M. Ericsson's Stockholm workshops the following year*

followed. Within a decade handset telephones were widely used throughout Europe and were popular with telephone users.

Before the development of the granular transmitter almost all the types of transmitters available were used in handsets. All were adversely affected by vibration to a lesser or greater extent and the granular transmitter was a considerable advance. Not only was it undamaged by vibration, a certain

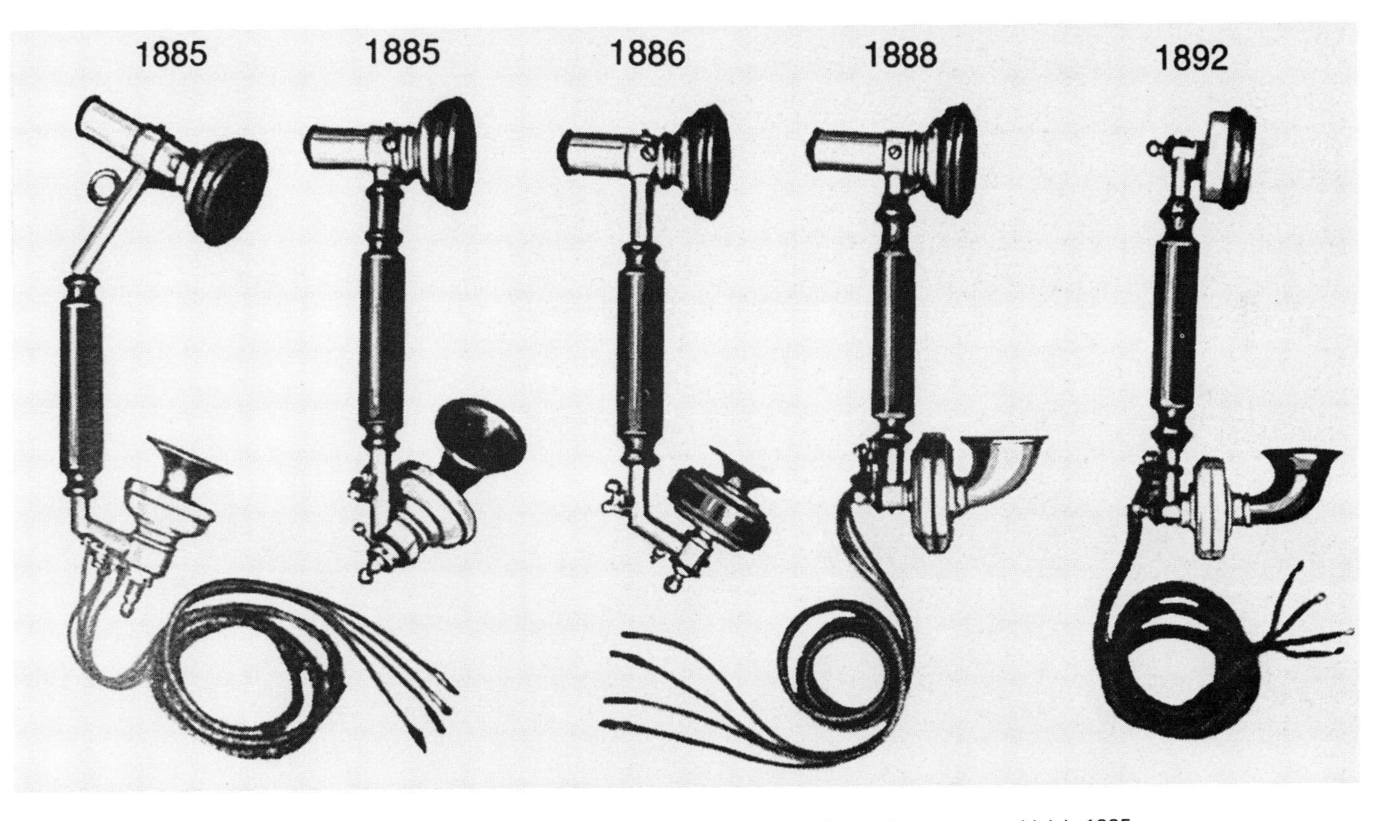

Fig. 3.2 *L.M. Ericsson handsets from 1885 to 1892. The bulky receiver magnet, which in 1885 protruded from the back of the receiver was, by 1892, contained within the receiver capsule. This was achieved partly by using better magnetic alloys and partly by more compact design*

Fig. 3.3 *L.M. Ericsson table telephone with handset, 1892. Reminiscent of the old Singer sewing machine and the brass knobs on a Victorian bedstead. This telephone was one of the most popular models ever produced. It was not only used in most European countries, including Russia, but also as far afield as South Africa, Australia and New Zealand*

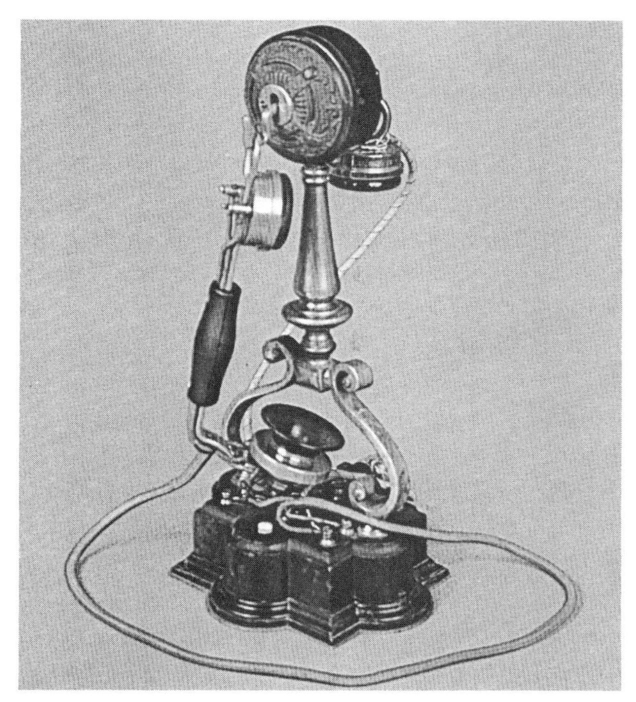

Fig. 3.4 *Telephone of 1893 with Berthon transmitter and Ader receivers made by the Société Générale des Téléphones. The wiring between the base of the telephone and the gravity switch-hook is concealed within the hollow column*

Fig. 3.5 *French Berthon-Ader telephone of the early 1890s on a 50 franc stamp of Senegal which was one of a series issued to celebrate the centenary of the International Telecommunications Union in 1965*

amount of vibration and mobility was actually beneficial and helped prevent packing.

In addition to the transmitters which played a major role in telephone development, others were invented that were never adopted on a large scale. One curiosity was the water-jet transmitter invented by Dr. Chichester Bell, who was Alexander Graham Bell's cousin. It was based on his researches which were the subject of a paper he read before the Royal Society in 1886. In the transmitter, a thin stream of water issued from a downward pointing tube and fell upon a platinum contact. On striking this contact, the water sprayed out horizontally in all directions and struck a platinum ring surrounding the contact. When an electrical current was passed from the contact to the ring via the spray, it was found that a minute deflection of the jet caused a considerable variation in the

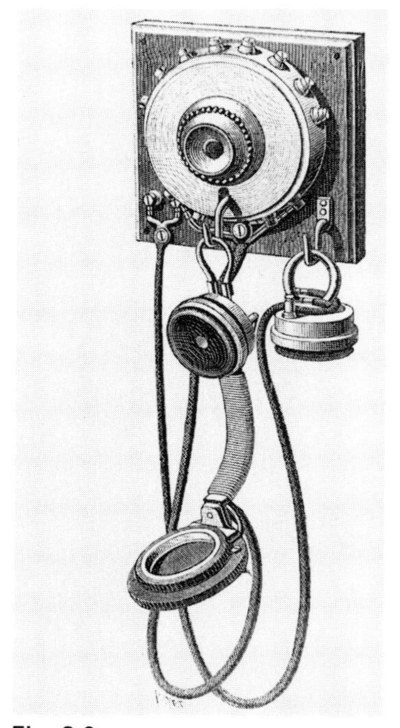

Fig. 3.6

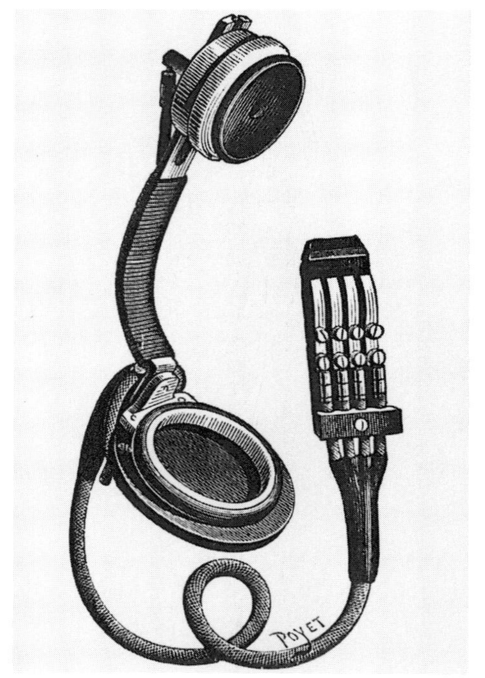

Fig. 3.7

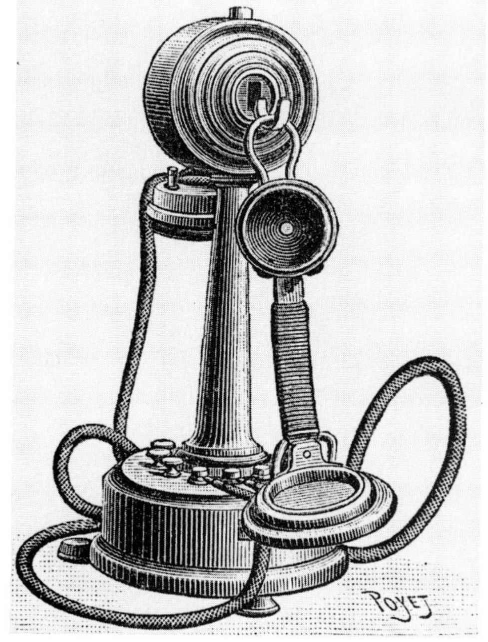

Fig. 3.8

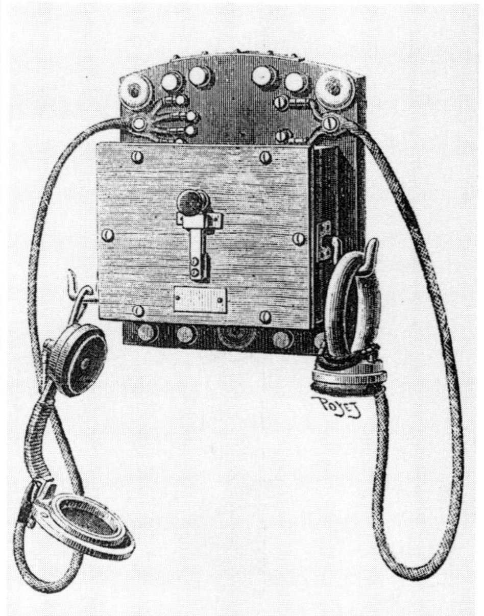

Fig. 3.9

Some handsets and handset telephones of the nineteenth century

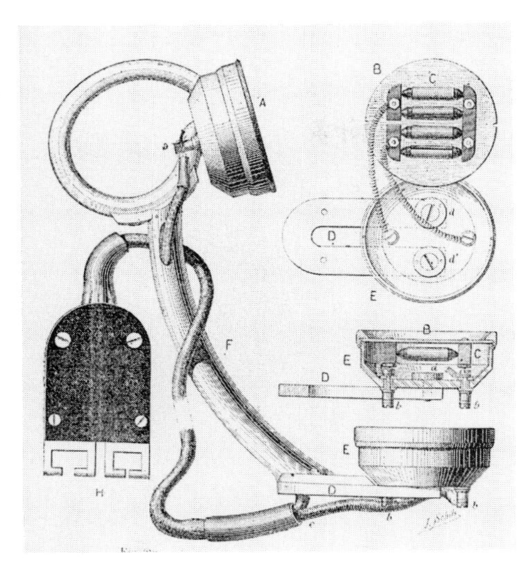

Fig. 3.10

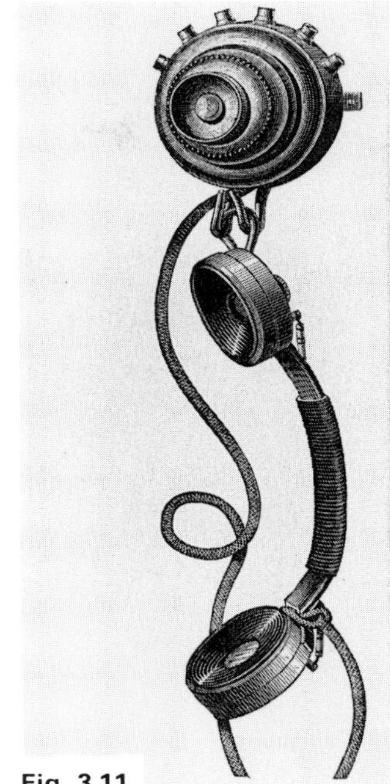

Fig. 3.11

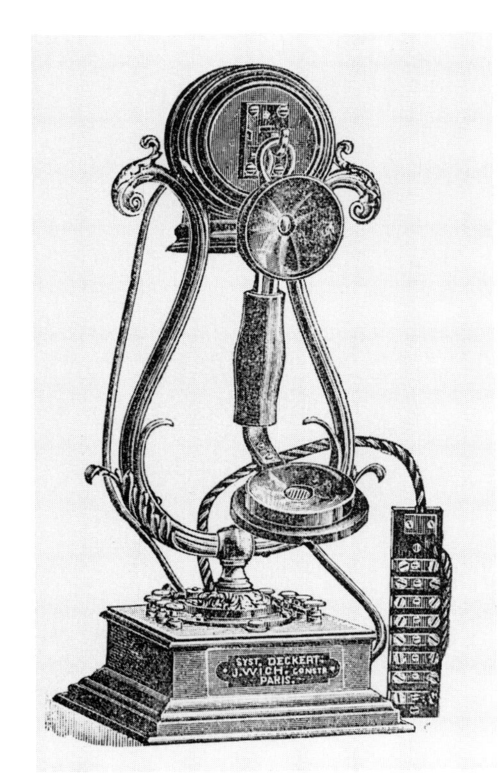

Fig. 3.12

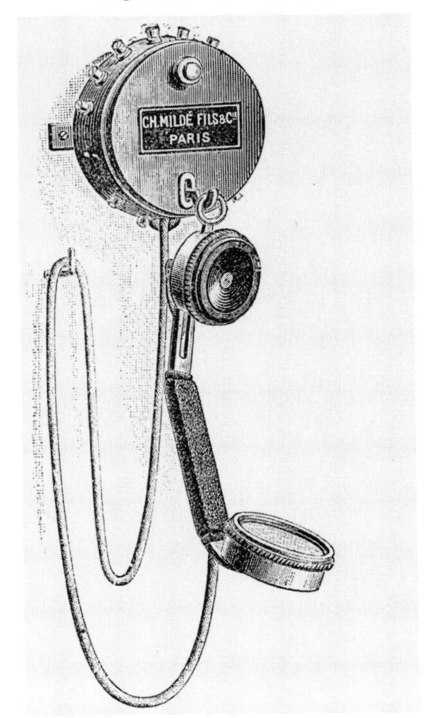

Fig. 3.13

Fig. 3.14 *Telephone incorporating the water-jet transmitter invented by Dr. Chichester Bell in the late 1880s. The transmitter is contained within a glass capsule at the bottom of the instrument. The circular object in the centre of the picture is the pump which is operated by the lever protruding on the right*

electrical current. Chichester Bell utilised this principle in his transmitter by causing the jet to be deflected by the impact of the sound waves. He subsequently built a complete telephone which was about three feet high and wall mounted. At its top, a tank acted as a reservoir for the water supply and at its base, a second tank acted as a sump to receive the used water. Water could be transferred from the sump back into the reservoir by a lever-operated pump. This lever was mechanically linked with the switch-hook mechanism so that it was impossible to make a call without operating the pump. A single operation of the lever filled the reservoir which then contained sufficient water for a seven minute telephone

Fig. 3.15 *Designed for call-offices, the cabinet telephone was introduced by the Bell company in the USA in the late 1880s. Like the duck-billed platypus, it is hard to classify. Is it a telephone or should it be called 'furniture'?*

conversation. If the call lasted longer, the pump could again be operated while the call was in progress.

The water-jet transmitter was never adopted on a large scale but interest in it was revived in the early years of the twentieth century when attempts were being made to send speech by radio. With the state of knowledge at that time, the only way to do this was to place a telephone transmitter in the output of the radio transmitter. However, when this was done, the output was restricted by the power-handling capacity of the telephone transmitter, so experiments were made with water-jet transmitters which could handle greater power. Although these

Fig. 3.16 *Wall telephone in widespread use around 1900, made by L.M. Ericsson of Sweden. Above the bell gongs is the protector with a moveable peg in the position it would occupy during a thunderstorm*

Fig. 3.17 *A desk telephone of about 1900 with Fitzgerald transmitter. Used for internal communication in office and home*

Fig. 3.18 *Compared with their European counterparts, most nineteenth century American telephones were utilitarian and functional. However, with appearance in mind, this Bell Company model of 1897 was cast in bronze. One interesting feature is the adjustable stem on which the transmitter is mounted. It can not only be raised and lowered but also rotated. When used in the completely reversed position the receiver rest was on the right hand side – a convenience for people who liked to hold the receiver in their right hand*

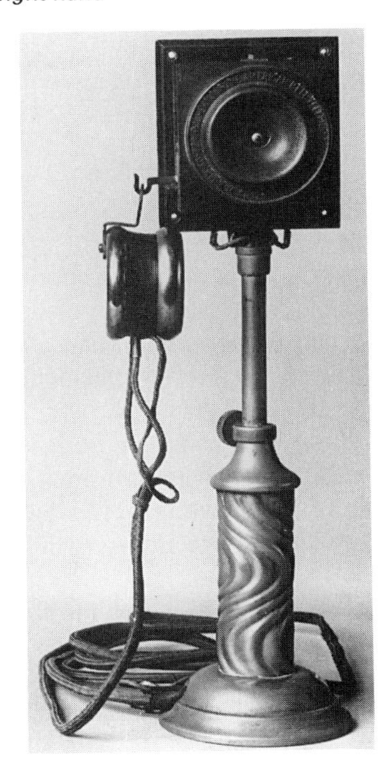

Fig. 3.19 *The influence of the artistic movement which became art noveau is shown in the 'spaghetti' type decoration on the column of this American Bell telephone of 1892*

Fig. 3.20 *The 'coffee mill' telephone made by L.M. Ericsson of Sweden in 1893*

experiments met with a degree of success the development of electronic valves enabled more efficient methods to be adopted.

The first public telephones were ordinary telephone instruments installed in 'attended call offices'. These offices generally consisted of a room with an attendant to collect the charge for the call and, if necessary, assist the caller in operating the telephone. The first telephones specifically designed for call office use were brought into service by the Bell company in the USA in the late 1880s. They were called 'cabinet' telephones and were built in the form of an ornately carved writing desk. The transmitter was mounted on the desk top and the receiver was hung on the left hand side while, in the lower part, there was a space for the batteries. All the rest of the components were in full view behind a bevelled plate glass window.

Cabinet telephones were manufactured over a number of years and, although the basic shape remained unchanged, there were alterations to the carving and general style. The transmitters were also unchanged. The earliest was the long distance transmitter with horizontal diaphragm. Later, more modern types were used. Although designed for call office use, the cabinet telephone was shown in

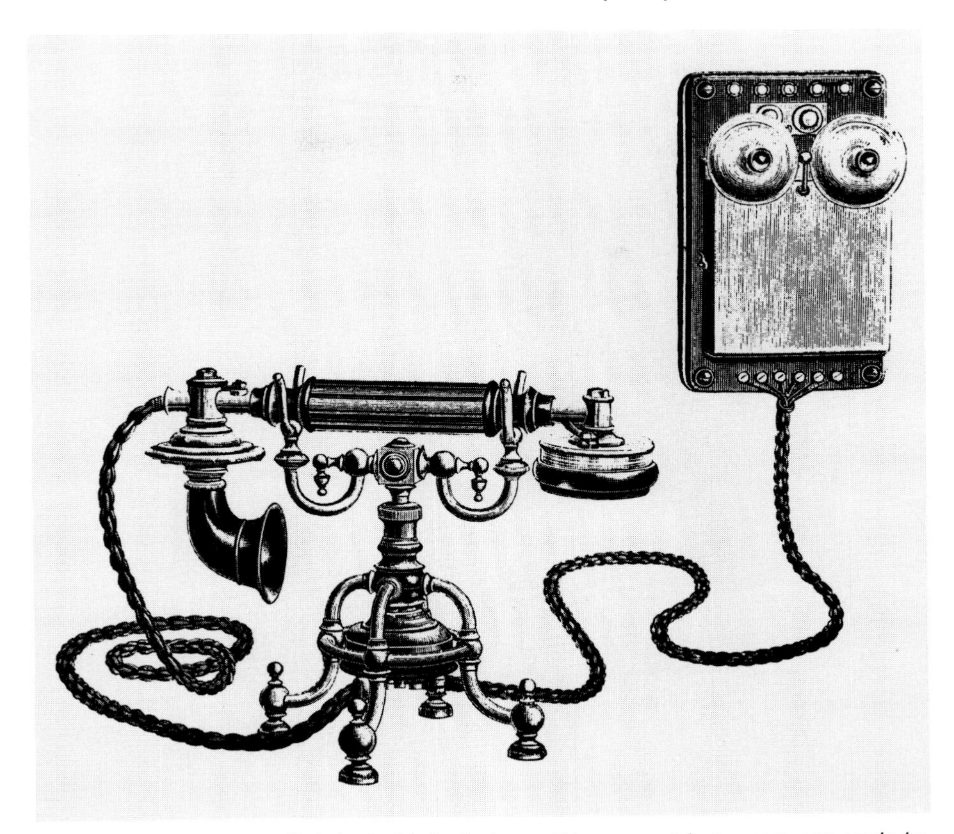

Fig. 3.21 *Sometimes called the 'spider' telephone, this compact instrument was made by L.M. Ericsson of Sweden in the 1890s*

the Western Electric Company's catalogue and may also have been used by ordinary subscribers.

In the early years of the telephone, when almost all lines were overhead, the risk from lightning was much greater than it is today. Many telephones incorporated simple lightning protectors intended to dissipate any excessive electrical discharge on the line. In most cases these protectors consisted of a number of metal plates screwed to the woodwork of the telephone or some other insulating material. A typical protector might have three plates, one of which was connected to earth while the plates adjacent to it were connected to each of the wires of the telephone line. The edges of the plates were positioned so that they were separated by minute gaps. Excessive voltage would cause sparks to jump these gaps, dissipating the electrical charge. To facilitate sparking, the edges of the plates were shaped like the teeth of a saw, with the points of the teeth on adjacent plates facing each other.

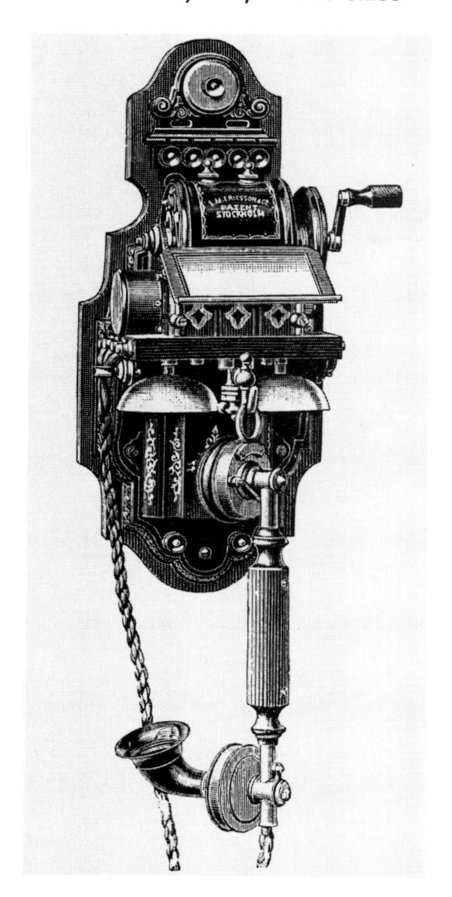

Fig. 3.22 *Characteristic of the 'telephones without cases' produced by L.M. Ericsson, this wall instrument, advertised in the Company's 1902 catalogue, had no front, top, bottom or side panels*

Fig. 3.23 *With carved and turned wooden case this German telephone of the 1890s is reminiscent of a cuckoo clock*

Fig. 3.24 *Without its handset this Siemens and Halske telephone of 1897 could be mistaken for a musical box*

Fig. 3.25 *Bavarian magneto telephone of 1893 with transmitter mounted on 'lazy tongs'*

Fig. 3.26 *Apparently inspired by a Greek or Roman urn, this Siemens and Halske telephone was, in fact, made in the 1890s*

As a further precaution, some protectors also had a metal peg which was normally parked in a hole in one of the plates but could be put into the gap between the plates. This enabled the subscriber to connect the line to earth when he felt it necessary.

With the passage of time, experience was gained and it was realised that the most effective place for a protector was at the point where the lines entered the building. So, from the 1880s onwards, protectors were developed as separate units and no longer formed part of the telephone instrument.

By the 1890s the ideas of telephone designers were crystallising and the candlestick telephone – one of the classic telephone shapes – started to become dominant. In its various forms it was destined to remain in use for over half a century and to become one of the most popular telephones of all time.

During the 1890s and early 1900s many unusual telephones were produced. In Sweden, L.M. Ericsson made one which is regarded as a classic. It was a table instrument with a cylindrical body and it became known as the 'coffee mill' telephone. The shape of the body limited the space for the magneto-generator and the larger, more powerful, generators could not be used. Generators which were small enough to fit into the body produced less power, and this restricted the

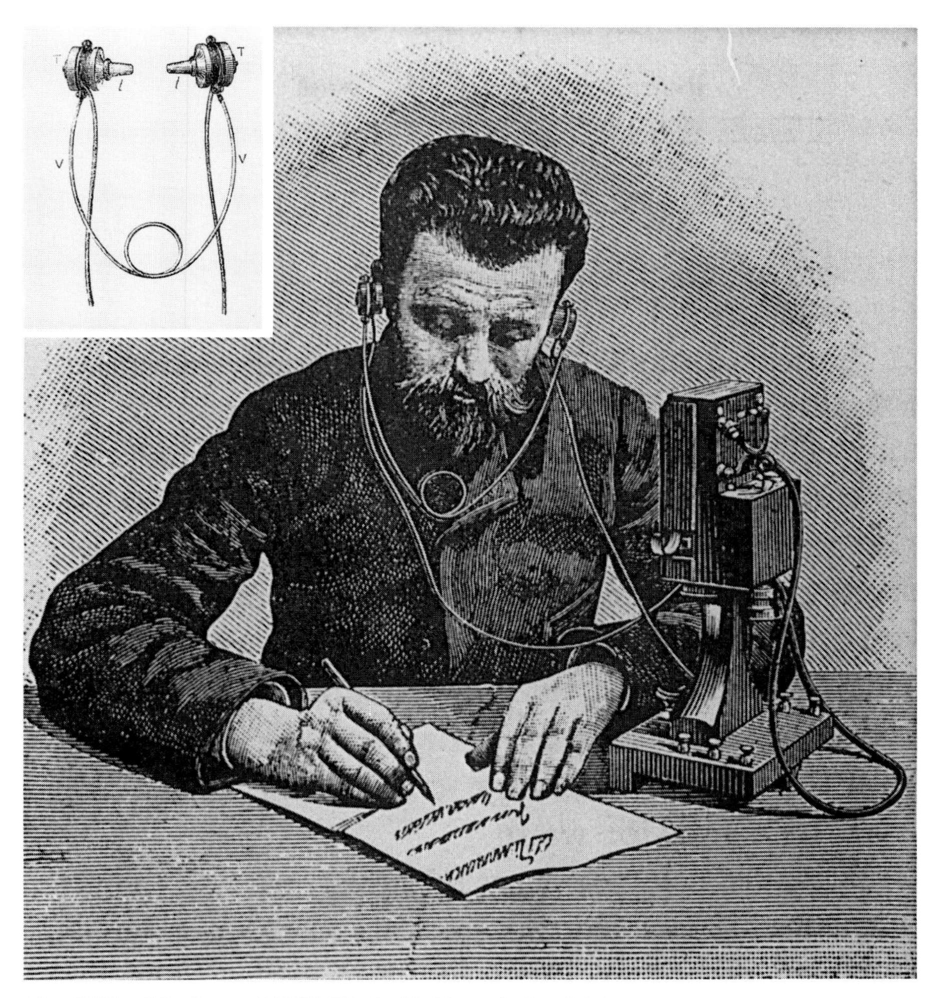

Fig. 3.27 *Telephone of 1890 fitted with Mercadier's miniature receivers*

circumstances in which the telephone could be used. Very few telephones of this type were made and they are now extremely rare.

Another unusual Ericsson instrument consisted of a handset with its associated cradle and gravity switch mounted on four curved legs to form a compact table telephone. All the other components were contained in a bell box screwed to the wall. The idea of separating the parts which needed to be in the telephone instrument, from those which did not, in order to make a compact table telephone, re-occurs in the designs of various manufacturers extending over many years.

A third instrument produced by L.M. Ericsson at this time was characteristic of the telephones without cases designed by this manufacturer. It was wall-mounted with the components screwed to a backboard, but it had no front, top, bottom, or side panels. The moving parts of the exposed magneto might well have trapped fingers or hair if safety guards had not been fitted. In the L.M. Ericsson catalogue it was claimed that the instrument was 'intended for use in the tropics', and that 'the employment of all wood has been carefully avoided as it very soon gets destroyed by ants and other insects'. This may explain why wood was not used. It does not explain why the case was dispensed with altogether. A metal case would have been insect-proof. A more probable reason for Ericsson's case-less telephones is the Company's pride in its workmanship and the quality of its components. Perhaps it felt that, like Victorian ladies' lace petticoats, they were too good to be forever hidden.

While some Swedish telephones of the 1890s had little or no case, there was quite a different approach to telephone design in most other countries. Across the Baltic, in Germany, telephones were quite different. Components were concealed, unless it was absolutely necessary that they should be seen, and in some cases it was almost as if designers were attempting to make instruments look as unlike telephones as possible. The German manufacturer, Mix and Genest, produced a telephone in a

Fig. 3.28 *Tall as an average person, this American floor-standing instrument, used in the 1880s and 1890s, was known as the 'tombstone' telephone*

Fig. 3.29 *The 'horsecollar' telephone enabled conversations to be conducted in absolute privacy; but special provision had to be made for telephone users to breathe!*

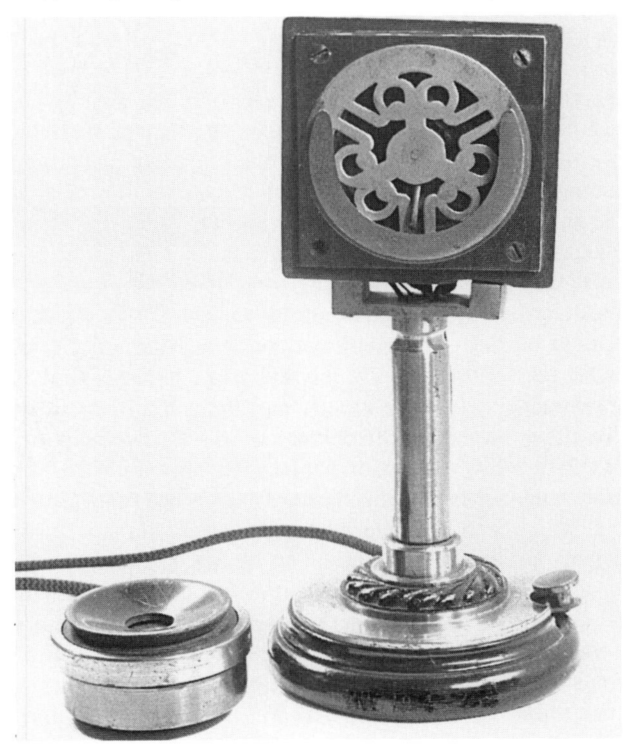

Fig. 3.30 *Stanley telephone for internal use with Byng granular transmitter. Made by the General Electric Company of London in the 1880s. When not in use, the earpiece was supported in the horseshoe-shaped cradle on the front of the transmitter*

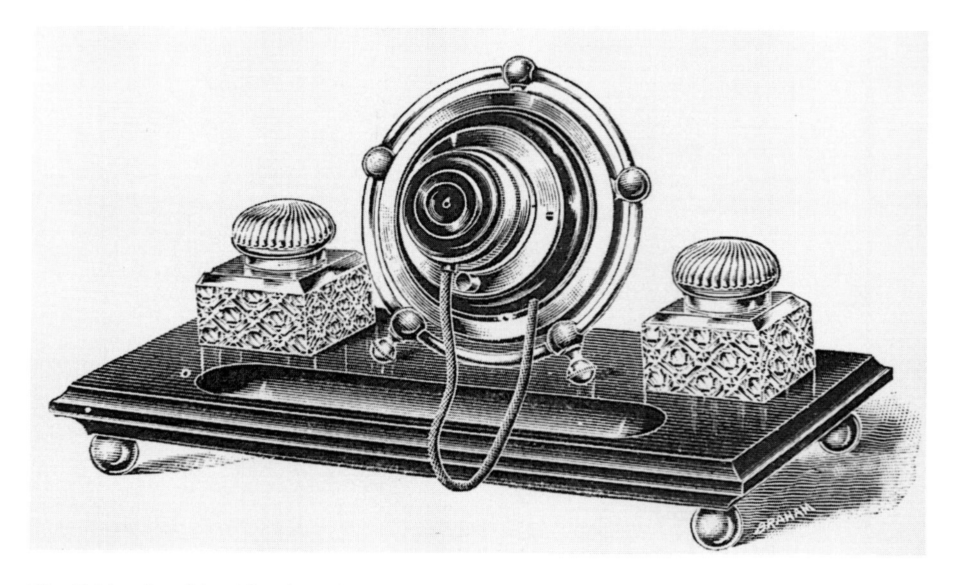

Fig. 3.31 *Combined Stanley telephone and inkstand of the 1890s which would grace the desk of any Lord or Lady*

carved wooden case, with turned wooden attachments, which was not unlike a cuckoo clock. But it was the telephone designs of Siemens and Halske of Berlin which were most influenced by this fashion for camouflaging telephones. Some of this firm's instruments were in the style of the carved stonework in classical architecture. One design in particular, available in either painted metal or polished bronze, could well have been copied from a Greek or Roman urn.

In France, one very unusual telephone resulted from the work of M. Mercadier who conducted lengthy experiments to determine the most effective dimensions for the component parts of the telephone receiver. Mercadier's conclusion, like that of Bell who had conducted similar experiments, was that it was not the individual dimensions, as such, which mattered but the ratio between the dimensions. In other words, the receiver could be scaled up or scaled down without detracting from its performance. By applying this knowledge, he produced a miniaturised telephone receiver weighing less than one and three-quarter ounces which was capable of reproducing sound without any loss of volume. This tiny receiver was designed to fit in the channel of the ear like a modern hearing aid. Two receivers were attached to a spring which held them against the ears as with a doctor's stethoscope. The advantage claimed for this arrangement was that it left both hands free for taking notes etc.

Even in America, where telephones were predominantly utilitarian, at least one unusual telephone was produced. It was floor-standing and as tall as an

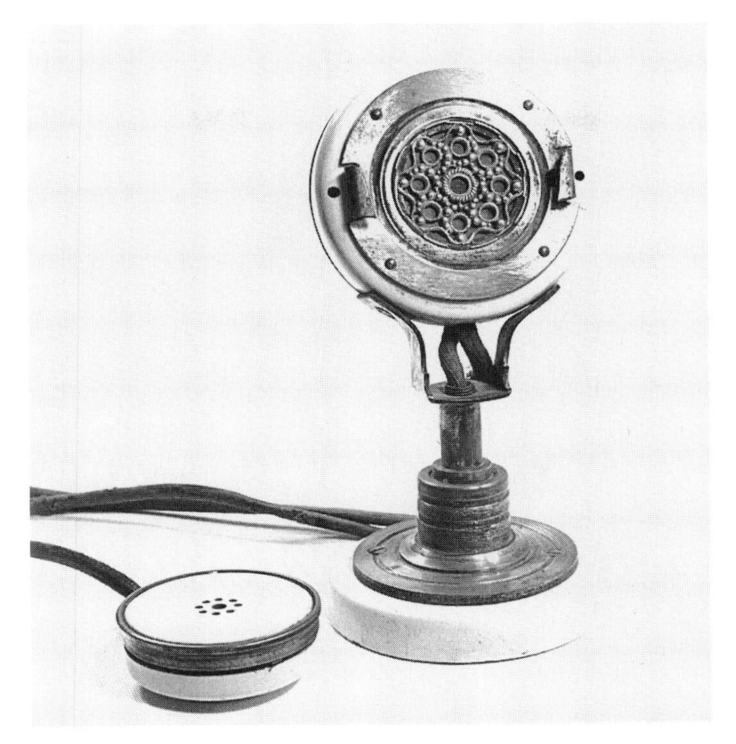

Fig. 3.32 *Pedestal telephone of the 1890s designed by George Lee Anders and A.J. Wilson, junior of London. The small brackets, on either side of the transmitter, support the receiver when the telephone is not in use*

average person. The size and shape of its wooden case made it look like a grandfather clock but it was commonly called the 'tombstone' telephone.

By far the strangest of all the telephones produced during the last decade of the nineteenth century was made in England. It was intended to stop bystanders hearing what a person using the telephone was saying. To achieve this, the transmitter was mounted in a wooden box with an elliptical opening into which the caller pushed his face. Soft material cushioned the edge of the opening and formed a seal which was sound-proof and, incidentally, airtight. In early versions of the telephone this cushion was faced with leather and stuffed with horse hair. Its appearance gave the instrument its popular name – the 'horse-collar' telephone, because it looked like the collars worn by cart horses. In later versions this cushion was replaced with an inflated rubber ring.

To enable the telephone user to breathe, a way had to be found to allow air to enter the box without letting sound out and a special ventilator was fitted. This had a series of glass baffles around which the air passed. The telephone was understandably unpopular and by 1900 it had faded into obscurity.

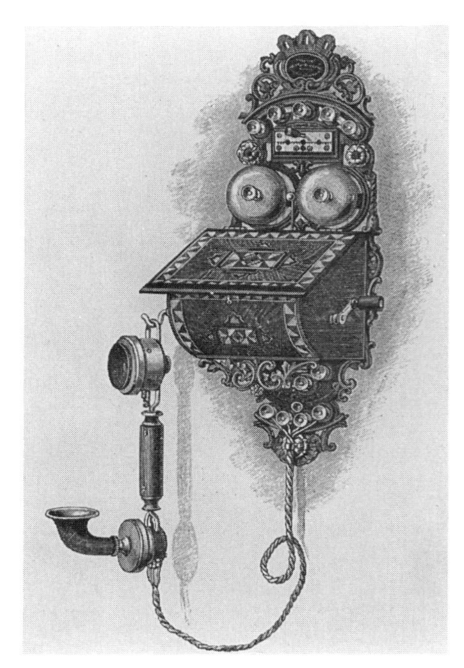

Fig. 3.33 *This handsome telephone of the 1890s was made by the Norske Elektrisk Bureau of Kristiania (now Oslo), Norway. The writing desk is made of sheet iron but is enamelled to give it the appearance of marquetry*

Fig. 3.34 *Telephone used by the Vienna Private Telegraph Company in the 1890s. The receivers were hung by their cords from switch-hooks. This practice was both inconvenient and caused unnecessary cord wear*

Some years later this telephone was displayed, with a number of other historical instruments, at an exhibition where it was noticed by the cartoonist Lowe. He sketched it and added the caption – 'Horse-collar transmitter, inventor deservedly unknown, probably suffocated.' What more could be said?

In Britain, a series of telephones were made around 1900 with the common feature that, when not in use, the receiver earpiece was hung over the front of the transmitter. One telephone of this type was designed by George Lee Anders and A.J. Wilson, junior. Other designs were manufactured by the General Electric Company and sold as 'Stanley' telephones. They were very simple instruments unsuitable for connection to the public telephone system. Their intended use was as internal, mainly domestic, telephones. In the large houses of the period, with their armies of servants, they would have proved their worth enabling the master or mistress to direct butlers, cooks, gardeners and grooms etc.

Chapter 4

The Electrophone

Today, most people think of the telephone as a means of conveying conversation. But, until the development of radio broadcasting, it was frequently used for the transmission of music, other forms of entertainment and, what in Victorian days was popularly called 'edification'. Indeed, it has been said that the telephone was used for music before it was 'able to talk'. This is true because the telephone of Reis is known to have transmitted recognisable tunes.

When Bell invented his telephone he gave a number of lectures and public demonstrations. In almost all of these he showed the telephone's versatility, including its ability to transmit music. When, for example, he demonstrated his telephone to Queen Victoria he prepared a programme which included a singer and also music played by a band. Similar experiments were made in many other countries.

One of the first attempts to broadcast music to a dispersed audience was made by Messrs. Moseley and Sons, a firm which provided private telephone lines for customers in Manchester, England. In order that the music might reach the maximum number of listeners, with reasonable loudness, a transmitter producing the greatest possible output was required. Such a transmitter was designed by the head of the firm's construction department, Mr. Alexander Marr. It was used for the transmission of music from Manchester theatres in 1880 and 1881. The music was received on ordinary telephone instruments.

In Paris, a complete system called the Théâtrophone was developed. It was first exhibited at l'éxposition d'électricité in 1881. Coin-in-the-slot machines were installed in hotels, clubs and other places where, for the expenditure of a franc, the public could listen for ten minutes to a live performance from a Paris theatre. An indicator on the side of the machine showed the theatre from which the performance was being relayed. This indicator was operated, by means of a separate telegraph circuit, from the Théâtrophone exchange in the Rue Louis-le-Grand which also distributed the music. At the theatre, two sets of telephone transmitters were used. One was on the left of the stage and the other on the right. The output from these transmitters was carried by separate lines and connected to separate earpieces at the receiving instruments. In this way a stereophonic

Fig. 4.1 *New Year, 1892. Partygoers listen to greetings by telephone*

effect was achieved. In the words of a contemporary account, 'The effect of a singer crossing the stage is very curious and realistic, and it is difficult to imagine that the singer is not in front of the listener.' Another commentator wrote, 'So the Théâtrophone is here. What is still lacking is the "Telephote" which would also enable us to see the actors or singers at all the distances over which they can be heard.'

Music was also transmitted in Munich, Vienna and Frankfurt.

In Britain, the National Telephone Company conducted one of the first inter-city experiments when, in 1891, theatrical performances in London were successfully transmitted to Birmingham. The following year, programmes were relayed from theatres in Birmingham, Manchester and Liverpool to a concert room at the Crystal Palace, Sydenham, near London. Visitors were charged three pence before 8.0 pm and six pence after 8.0 pm to listen to the programme for ten minutes. During the six months that the concert room was open, almost 60,000 people attended.

The demonstration of public interest in the Sydenham experiment led to the creation of a permanent service. In 1894 the Electrophone Company was formed – 'Electrophone' being the name of the British version of the French Théâtrophone. This company worked in close association with the National Telephone Company – then the principal provider of telephone service in the United Kingdom. Starting in 1895, subscribers were offered a selection of music,

Fig. 4.2 *Coin-in-the-slot Théâtrophone receiver*

Fig. 4.3 *The Théâtrophone being used in a Paris hotel, 1892*

Fig. 4.4 *Poster designed by Jules Chéret for the Compagnie du Théâtrophone*

theatrical performances, public speeches and church services for an annual rental of £5.

Electrophone subscribers were provided with an 'intermediate switch' which enabled them to connect their telephone lines to either their telephone or the Electrophone apparatus. The Electrophone apparatus itself generally consisted of a small table, rather like a plant stand, to which a number of Electrophone receivers were connected. Each receiver consisted of two earpieces supported on a horseshoe shaped metal band joined to a handle. This arrangement enabled the earpieces to be held to the ears for long periods without unduly tiring the arms or disarranging the hair.

When a subscriber wished to use the Electrophone he called the exchange using his normal telephone and asked for 'Electrophone service'. He was then connected to the Electrophone operator who could offer him a selection from up to thirty programmes. The subscriber then used his intermediate switch to bring the Electrophone apparatus into circuit.

Both Queen Victoria and King Edward were Electrophone listeners, and a catalogue of the actors, singers and musicians who, at one time or another, performed to the Electrophone, would read like a 'Who's Who' of Victorian and

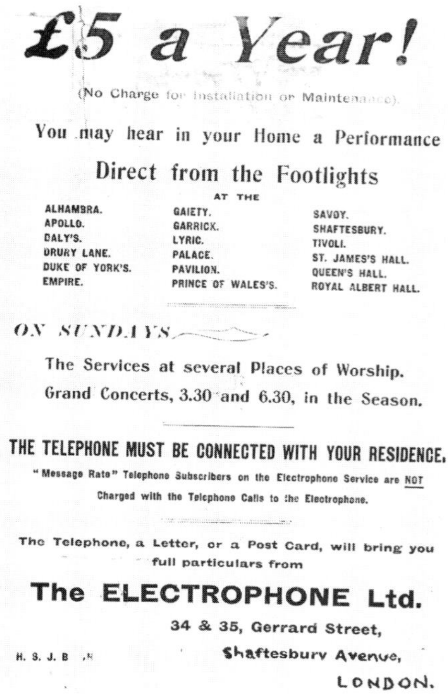

£5 a Year!

(No Charge for Installation or Maintenance).

You may hear in your Home a Performance

Direct from the Footlights

AT THE

ALHAMBRA.
APOLLO.
DALY'S.
DRURY LANE.
DUKE OF YORK'S.
EMPIRE.
GAIETY.
GARRICK.
LYRIC.
PALACE.
PAVILION.
PRINCE OF WALES'S.
SAVOY.
SHAFTESBURY.
TIVOLI.
ST. JAMES'S HALL.
QUEEN'S HALL.
ROYAL ALBERT HALL.

ON SUNDAYS,

The Services at several Places of Worship.

Grand Concerts, 3.30 and 6.30, in the Season.

THE TELEPHONE MUST BE CONNECTED WITH YOUR RESIDENCE.

"Message Rate" Telephone Subscribers on the Electrophone Service are NOT Charged with the Telephone Calls to the Electrophone.

The Telephone, a Letter, or a Post Card, will bring you full particulars from

The ELECTROPHONE Ltd.

34 & 35, Gerrard Street,

H. S. J. B ,N

Shaftesbury Avenue,

LONDON.

Fig. 4.5 *Although you could not see the Gaiety Girls with the Electrophone, you could hear them; and plays, concerts and church services too. All this was offered for £5 a year in this 1902 advertisement*

Fig. 4.6 *1902 advertisement offering 'A theatre at home', and for 'under 2d per day'*

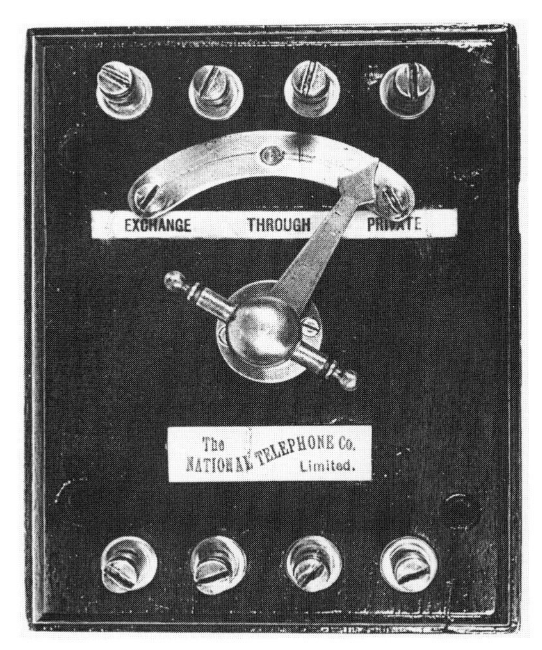

Fig. 4.7 *The 'intermediate through' switch was originally developed to extend a call from a main telephone to an extension. It was also used to bring into circuit Electrophone apparatus*

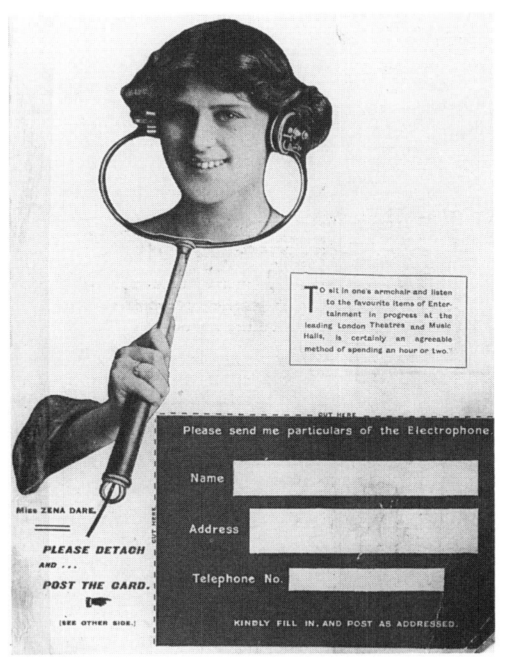

Fig. 4.8 *1902 advertisement showing the 'headset on a stick' used by Electrophone subscribers was sometimes called the 'lorgnette' headset. It enabled Electrophone receivers to be held to the ears with the minimum effort and without disarranging the hair*

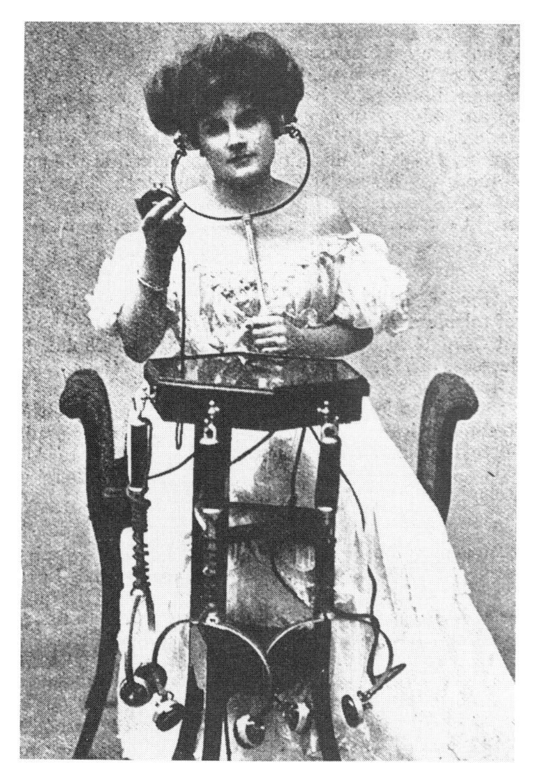

Fig. 4.9 *Many Electrophone subscribers were provided with a special table to which a number of Electrophone receivers were attached. In this 1902 photograph the lady is shown holding a telephone transmitter. This enabled her to speak to the Electrophone operator in order, for example, to ask for a different programme*

Edwardian entertainment. But not all the acts were an Electrophone success. A member of the National Telephone Company staff wrote in 1907 'It is essential to have a good variety of both instrumental and vocal music – people don't want to listen to performing dogs or elephants.'

The Electrophone service achieved a remarkable success when, in 1909, it relayed a political speech from Glasgow to London – the first international relay. In 1912 the National Telephone Company was taken over by the British Post Office but the Electrophone continued to function, collaborating with the Post Office as it had previously done with the Company. The following year, Electrophone subscribers were able to hear Faust transmitted direct from the Paris Opera, while Théâtrophone subscribers in Paris were able to hear excerpts from Tosca relayed from the Royal Opera House, Covent Garden.

The achievements may seem commonplace now but it must be remembered that the results were obtained without any form of amplification. It is little

Fig. 4.10 *In Edwardian London, ladies and gentlemen being entertained by the Electrophone*

wonder that the programme was sometimes faint and distorted although there may have been more than one reason for this. It is recorded that several of the choristers in the church of St. Anne's, Soho, London were reprimanded for secretly stuffing the loose sleeves of their surplices into the Electrophone transmitters 'to see what would happen'.

In Budapest, the telephone was used to disseminate news. This service, called 'Telefon-Hirmondo' commenced in 1893. At the central office the news was read by 'stentors', that is to say announcers with powerful voices. The company was operated on similar lines to a newspaper with its own editorial staff and reporters. There were even advertisements and a twelve second 'slot' cost one florin. The service commenced at 10.30 am and finished at 10.30 pm on each day except Sunday. At the peak of its popularity in 1901, Telefon-Hirmondo had over 6,000 subscribers.

The relaying of entertainment by telephone was never as popular in the USA as it was in Europe in spite of Bell's early demonstration of music transmission. However, there are reports of music being sent by telephone from 1893 onwards. One place where the idea was tried was Wilmington, Delaware. In 1909 a service with the grand-sounding name 'Tel-musici' was made available. It was, in reality, no more than an arrangement whereby a telephone operator would play a gramophone record on request. In addition to their normal telephone, Tel-musici subscribers were provided with a special loud-sounding receiver to which a

Fig. 4.11 *This picture of an Edwardian glamour girl, used in a 1902 advertisement for the Electrophone, was intended to catch the eye, but people really did change into evening dress to listen to a programme from a theatre, and on Sundays, when they listened to a religious service, they wore their 'church clothes'*

megaphone could be fitted. Tel-musici coin-operated boxes were also installed in restaurants, hotels etc. A directory listing the records available was provided and people wishing to hear a record asked for it by number. Seven cents per record was charged for opera and three cents for other music. There was a minimum yearly payment of 18 dollars.

After the 1914–1918 war there were plans in Britain to extend and improve the Electrophone service using valve amplifiers. However, with the beginning of radio broadcasting in the early 1920s there was a rapid decrease in the number of Electrophone subscribers. The service became uneconomical and finally closed on 3 June 1925 except for a few lines in the Bournemouth area which continued to be used until the mid 1930s.

Radio broadcasting benefited from the technology which had already been developed for the telephone and the Electrophone. The microphones used by the early broadcasters were ordinary telephone transmitters and the technique

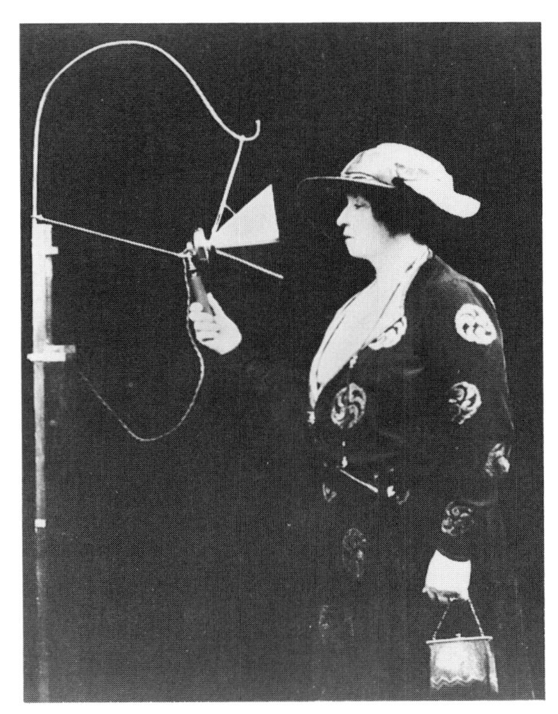

Fig. 4.12 *Dame Nellie Melba was the first great operatic celebrity to risk her musical reputation on the radio when, in 1920, she broadcast from Chelmsford, England. The experiment was a success and many people 'listened in'. Her willingness to take this risk may well have been partly because she had sung on the Electrophone from Covent Garden and been highly acclaimed. The microphone she used at Chelmsford was an ordinary telephone transmitter to which a cone was added to increase efficiency*

of positioning them had already been learned. As early as 1896 engineers positioning transmitters for the Electrophone were advised – 'The transmitters should not be fitted in close proximity to the bass drum or the trombone of the orchestra.'

In the Netherlands and Switzerland, arrangements were made to relay radio programmes over telephone lines and this service remained in use until well after the Second World War.

Chapter 5

Special purpose telephones

The Théâtrophone and the Electrophone were developed during the two decades following the invention of the telephone. This was a period when the rapidity with which technology was advancing, was matched by the ingenuity with which new uses for the telephone were being found. These included street telephones for the police, the fire service and for public utilities such as tramways.

Even before 1890, telephones were to be found in the streets of some American cities. 'Alarm boxes', fitted in prominent positions, contained telephones connected to the local police station. To prevent misuse, these boxes were locked and the keys were given to what were called 'chief citizens', these presumably included shopkeepers and other responsible people. As an added precaution against misuse, after a key had been used to open a box, a special mechanism prevented it being removed from the lock except by a policeman. A serial number on the key enabled the owner to be identified.

Misuse was an ever-present possibility uppermost in the minds of street telephone designers. The problem was to devise a telephone which would provide the best possible service without being vulnerable to improper use or damage. An instrument which went a long way to meet this requirement was the 'Digeon' apparatus adopted in 1892, by the fire service, for the streets of Paris. It consisted of a cast-iron pillar having in its upper part a compartment with two doors. In case of emergency, the main door could be opened by breaking a pane of glass. Behind this door a telephone mouthpiece enabled the caller to speak directly to the fire station. The second, smaller door, was normally locked but could be opened with a key carried by the firemen. It revealed a socket into which a portable telephone handset could be plugged to make reports and receive information from the fire station.

Fire pillars were used in Britain too, but they were smaller and much simpler. They were basically electrical fire alarms, actuated by breaking a glass window and pulling a lever, but there was also a socket into which a fireman could plug a portable telephone.

Telephones were also installed for tramways. The instruments used were conventional telephones, modified to make them resistant to weather, and

Fig. 5.1 *A street telephone in Chicago in the 1880s*

mounted in lockable iron boxes. The importance of telephones for the efficient running of an extensive tramway network can be appreciated from the fact that at one time, in the relatively narrow streets around the Elephant and Castle in South London, no less than twelve tram tracks converged in an enormous cat's cradle of intersections. A continuous stream of trams moved over these tracks at all hours of the day and night and the chaos that could have resulted from an immobilised tram can well be imagined.

Street telephones were used by numerous tramways in various parts of the world, but it was very unusual to have telephones on the trams themselves. One place where this was done was the seaside town of Llandudno in North Wales. Between 1898 and 1902 the Great Orme Tramway Company built a line from Llandudno to a point 679 feet above sea level not far from the summit of the Great Orme. There was a minimum gradient of 1 in 3·6 and in places the line was extremely steep. It was thought to be too steep for electric traction and the cars were hauled by cables in a similar way to the cable cars of San Francisco. But the cars in Llandudno also had trolley booms running on an overhead wire which was used to provide a telephone circuit. This was fitted because it was considered necessary to have communication between the winding engine man, situated in the steam-powered engine house about halfway along the line, and the tram driver, or brakesman as he should correctly be called. The telephones themselves

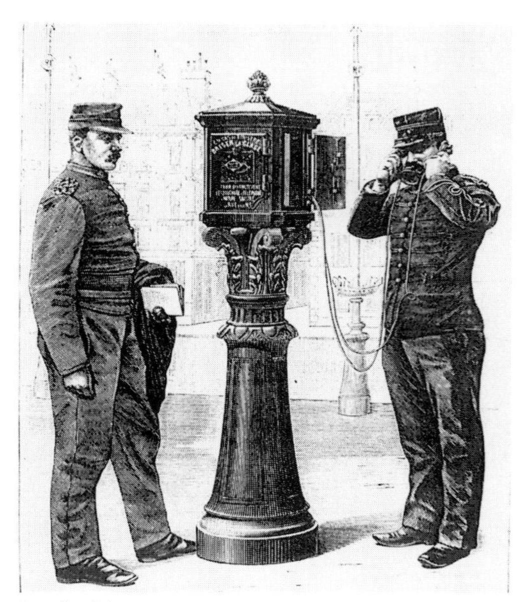

Fig. 5.2 *Fireman using the Digeon apparatus which was adopted for use in the streets of Paris in 1892*

Fig. 5.3 *Flower sellers near the Madelaine in Paris with Digeon apparatus on the left. c. 1925*

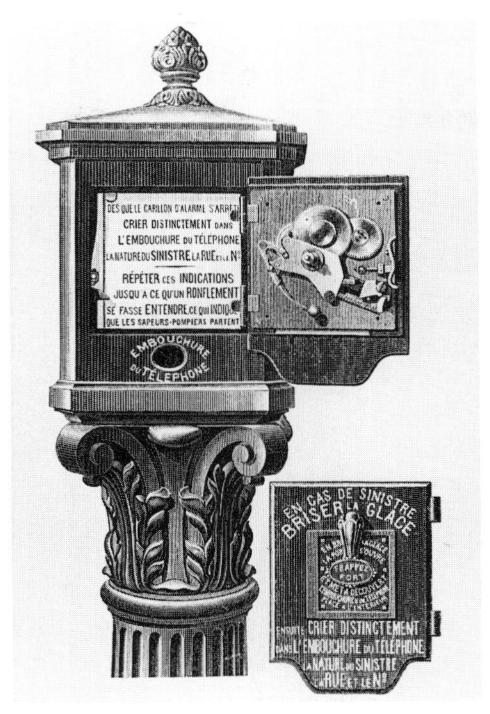

Fig. 5.4 *Detailed view of the Digeon apparatus showing the door used by the public*

were quite ordinary wooden-cased wall type instruments of the period, mounted on the open platform of the tram, with no more than a canvas cover to protect them from the weather and sea air.

With the increasing demand for telephones in office and home there was a parallel demand for telephones in public places. As early as 1878 the first public telephones were installed in the USA. In Britain, by 1882, public telephones were in use in the Stock Exchange and the Baltic Exchange in London, as well as the Wool Exchange in Bradford. Standard telephones of the period were used and payment was made to an attendant. However, with Victorian enthusiasm for penny-in-the-slot machines it was inevitable that it would not be long before someone would devise a mechanism to collect the money automatically.

In the mid 1880s there were a number of coinbox patents. In England, boxes were installed in Manchester and many Lancashire towns. About the same time the 'Smith and Sinclair' coinbox, named after its inventors, was widely used in Glasgow, Scotland. These boxes had slots for both penny and sixpenny pieces and different signals were sent to the exchange depending on which coin was inserted. There was an unusual feature. In the front of the box there were two slots into which a special 'pass' or 'key' could be fitted. When the pass was turned

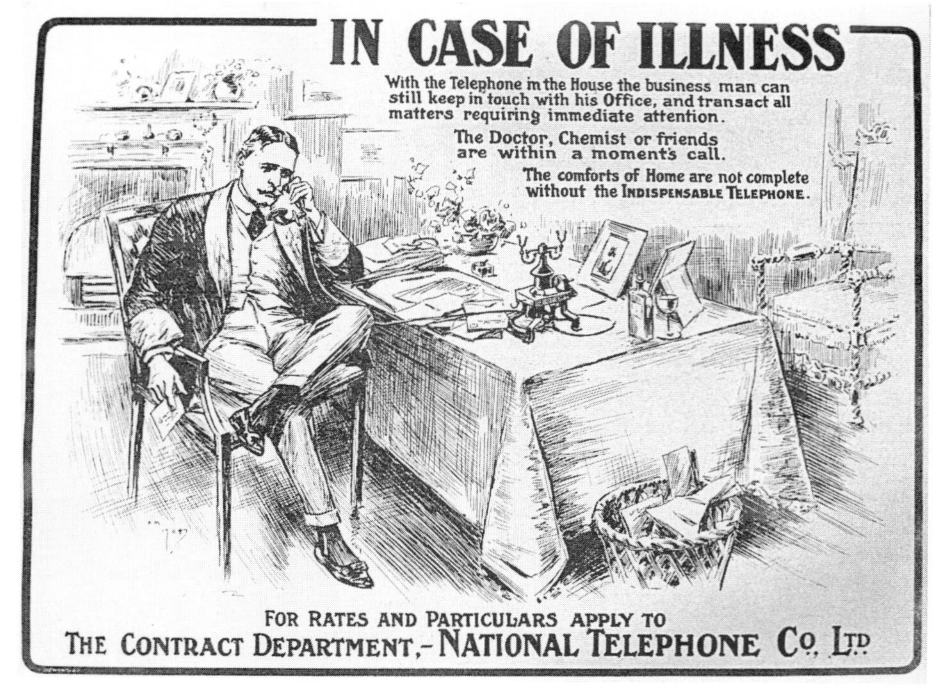

Fig. 5.5 *Advertisement of the early 1900s used in Britain by the National Telephone Company*

in the slot it had the same electrical effect as when a coin was inserted. This enabled the telephone company's subscribers to make calls without inserting money.

In the USA, coinbox patents date from 1885[18] and the first coinbox in America was installed in Hartford, Connecticut in 1889. Hartford became the home of the Gray Telephone Pay Station Company which manufactured an entire range of coinbox telephones used throughout North America. One of the Company's early models made an unusual concession to comfort. It was fitted with a metal elbow rest to support the forearm of the telephone user as he held the receiver to his ear. A telephone of this type was installed in Dawson City in the Yukon at the time of the gold rush in the late 1890s. It had coin slots for nickels (5 cents), dimes (10 cents), quarters, half dollars and silver dollars.

In most early European coinboxes, the insertion of the coin caused an electrical contact to be operated. This either completed the telephone circuit, so that a call could be made, or sent a signal to the exchange, to inform the operator that money had been inserted.

Coinboxes manufactured by the Gray Company in the USA, operated in a completely different way from their European counterparts. They had one or

Fig. 5.6 *Firemen using a fire pillar in Tottenham, London*

more gongs which were struck by a falling coin. The sound produced was heard by the telephone operator who connected the call. In the earliest Gray Company boxes, the sound of the gong was conveyed to the telephone transmitter via a hollow tube called a 'deflector'. But experience showed that this was unnecessary and that adequate sound was conducted to the transmitter through the metalwork of the instrument.

Almost from their inception, coinboxes were manufactured in two different types: those integrated with telephones, and independent coinboxes which were used with an ordinary telephone. With the Gray system, completely independent coinboxes were not possible because the sound of the gongs was carried by the physical link between the coinbox and the telephone transmitter. This led to the development of what was, in effect, an adapter or add-on unit which converted a normal telephone into a coinbox instrument. With the types of wooden wall telephone common in North America during the early 1900s, one way of making this modification was to remove the transmitter, fit a coinbox to the telephone backboard and then refit the transmitter on the front of the coinbox. Alternatively, wall telephones could be converted by mounting a coinbox alongside the telephone on a metal back-plate which extended behind the telephone backboard. One of the normal transmitter fixing screws was then

Fig. 5.7 *A tramway telephone in the streets of Glasgow, c. 1900*

replaced with a specially long screw which engaged a threaded hole in the back plate. By this means, vibrations from the gong in the coinbox were carried via the metal back-plate to the transmitter fixing screw and then to the transmitter, so as to be heard by the operator. A similar technique was used to convert candlestick telephones to coinbox use. As well as being used in the USA and Canada, Gray Company coinboxes were manufactured for other countries including Argentina, Australia and Brazil.

From the early days, many industries made use of telephones. Often ordinary instruments were adequate but some industries required special types of telephone. These special requirements might be simple: for example, equipment might need to be robust, to withstand rough usage, or water-tight and weather resistant. But there were also industries where telephones had to meet some very special needs. One such industry was coal mining. Since electrical equipment was first introduced into coal mines, a major problem has always been the sparks produced when electrical circuits are disconnected. Even the tiny movements of the carbon granules in a telephone transmitter produce minute sparks and, small as they are, they can ignite the explosive mixtures of methane gas and air frequently found in coal mines. In such an environment, an ordinary telephone

Fig. 5.8 *London taxi drivers of 1910 about to use a telephone fitted to a lamp standard near their rank*

could cause a disaster and special techniques had to be devised to overcome this danger. An early precaution, which has stood the test of time, was to seal the electrical components in an air-tight case, leading the wires in through air-tight glands. To enable the telephone user to listen and to speak, the diaphragms of the transmitter and receiver had to be exposed but gaskets were fitted round their edges in order to keep the telephone air-tight. Similar telephones were needed in oil storage depots, explosive factories and chemical plants where inflammable or explosive substances were processed.

Quite a different type of telephone was required by railway companies to enable them to control train movements. The places where railway telephones were needed were often far apart, and the cost of telephone lines was a major consideration. Systems which could be installed with minimum line costs were therefore particularly attractive. One such system was the 'omnibus circuit', that is, a single telephone line to which a number of telephones were connected. It was not unusual for railway companies to have omnibus circuits serving twenty or more telephones. When making calls, code ringing was used. Some of the coded rings were quite complicated. For example, 'two short rings, followed by a long ring, followed by two more short rings', or, 'three rings followed by a pause,

Fig. 5.9 *One of the trams on the Great Orme in Wales. These were probably the only trams on which telephones were fitted*

followed by four more rings'. Because it was easier to code ring by pressing a button than by turning a handle, direct current calling was often used in preference to magneto calling.

The principal disadvantage of omnibus circuits when used by the public is the lack of privacy but, when used by the railways, this did not matter. Railways used telephones for official business and lack of privacy was often considered to be an advantage. For example, in an emergency, telephone calls could be broken into and urgent messages sent.

The basic type of instrument used with railway omnibus circuits was the 'one-button' telephone, so called because it had one press button for code ringing. Another much used telephone was the two-button telephone. This was used when omnibus circuits were also connected to a switchboard. One of the telephone's buttons was used to call the switchboard and the other was used for code ringing between telephones.

A more sophisticated type of instrument, used almost as widely as the one and two button telephones, was the 'selective ringing' telephone. It enabled a number of different ringing conditions to be connected to a line while each telephone responded to one specific ringing condition only. It was particularly useful on busy omnibus circuits where, without such a telephone, constant ringing would distract staff from their work.

Fig. 5.10 *A 1902 telephone fitted on the open platform of a cable-hauled tram on the Great Orme near Llandudno, Wales, to enable the driver to speak to the man controlling the winding engine*

Another way of saving line costs was for the railway companies to make double use of their wires. Apart from the telephone, railways needed wires for sending telegraph messages and to operate signalling equipment. Both the low speed telegraphs used on railways and the railway signalling equipment, operated from slowly varying line currents. By contrast, telephones produce speech currents which vary relatively rapidly and do not affect telegraph or signalling equipment. If, therefore, a telephone was used that was unaffected by slowly varying currents, both types of equipment could be used simultaneously on the same wire. This was called 'superimposed working'.

A telephone with the necessary characteristics for superimposed working was the 'Phonopore'. Coils of wire wound in a special way were used in the receivers to make them insensitive to slowly varying currents. The quality and volume of speech was seldom perfect and the situations in which the telephones were used were often noisy. To help overcome these problems, Phonopore telephones were generally fitted with two receivers so that the telephone user could hold one to each ear. This improved reception and helped exclude unwanted noise. Because both magneto and direct current bells are operated by slowly varying currents which would interfere with telegraph or signalling equipment, normal methods of ringing were impossible. Phonopore telephones therefore used a buzzer to

Fig. 5.11 *Unusual street telephone of the early 1900s used by the police in St. Paul, Minnesota, USA. This telephone had two magneto-generator handles to call the switchboard at the police station. One, used when making routine reports, operated the normal calling signal. The other operated a special signal to indicate that the patrol wagon was needed*

generate the electrical equivalent of a high pitched note which was connected to the line as a calling signal. At the distant telephone, in addition to the two receivers used for conversation, a third was used solely to receive the calling signals. It took the place of the bell in a conventional telephone. This receiver was of the Collier-Marr type and was very sensitive and efficient. It responded to signals generated by the buzzers and emitted a high-pitched note through a small horn which acted as a megaphone. The volume was loud enough to attract the attention of anyone nearby. Several Phonopore telephones were often connected to a single telegraph line and code ringing was used.

Fig. 5.12 *Smith and Sinclair telephone coinbox installed in Glasgow in the mid 1880s. The slots on the top were for penny and sixpenny coins while the slots on the front were for a special 'pass' used by subscribers*

Another industrial problem, which was solved by designing special telephones, was encountered in electricity generating stations where there was a risk that the telephone circuits might come into contact with high voltage. This could occur as the result of a fault, either in the power station itself or, much more likely, on the lines between power stations. In the early days of power transmission, the telephone lines which provided communication between power stations were sometimes carried on the same supporting standards as the power lines themselves. If, due to a fault, the telephone circuits were exposed to high voltage, their modest insulation would do little to prevent a person using the telephone from getting a serious electric shock.

Fig. 5.13 *The front and reverse sides of a 'pass' issued by the National Telephone Company in the mid-1880s to fit the Smith and Sinclair coinbox*

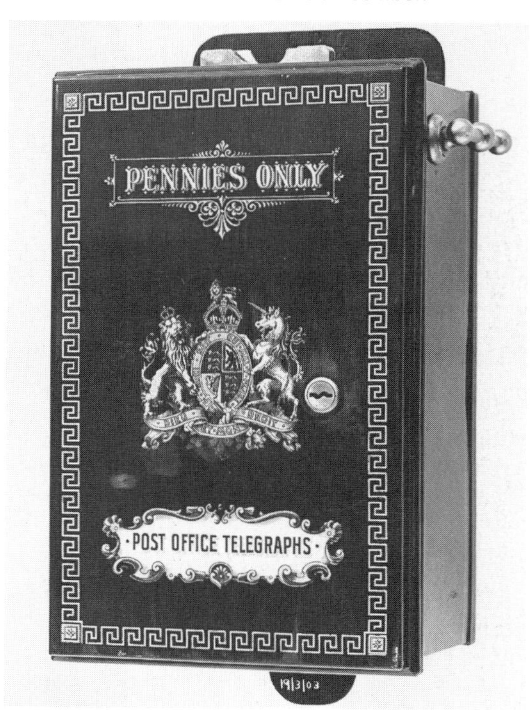

Fig. 5.14 *Coinbox of the early 1900s made by L.M. Ericsson. In its standard form the front cover was enamelled with the Ericsson trade mark but, if they wished, telephone administrations could have their own crest or logo. This box was used in Britain by the telegraph department of the Post Office which provided telephone service in a number of places. The Coat of Arms of the British monarch is enamelled on the box because, at that time, the Post Office was part of the Civil Service and nominally under the control of the Crown*

Fig. 5.15 *Coinbox telephone in black walnut made by the Gray Pay Station Company of Hartford, Connecticut, USA. The telephone has an elbow rest for the comfort of telephone users and five coin slots for nickels, dimes, quarters, half dollars and silver dollars. A telephone of this type was used in Dawson City in the Yukon during the gold rush of the late 1890s*

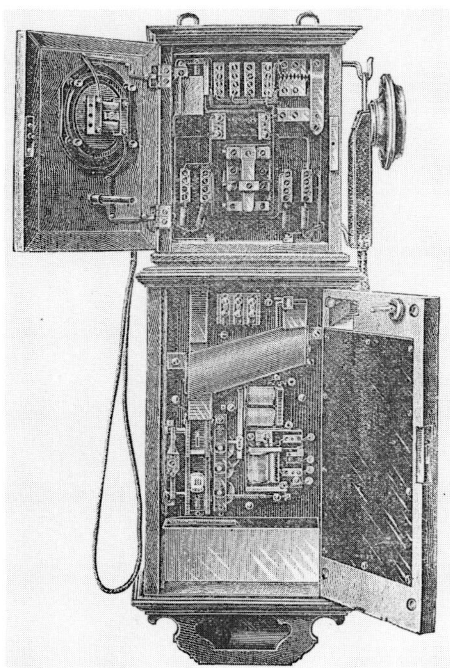

Fig. 5.16 *A German coinbox telephone of the late 1880s made by Mix and Genest. The coin was checked for thickness, diameter and weight and was rejected if incorrect. There was also provision to refund the coin if the wanted number was engaged*

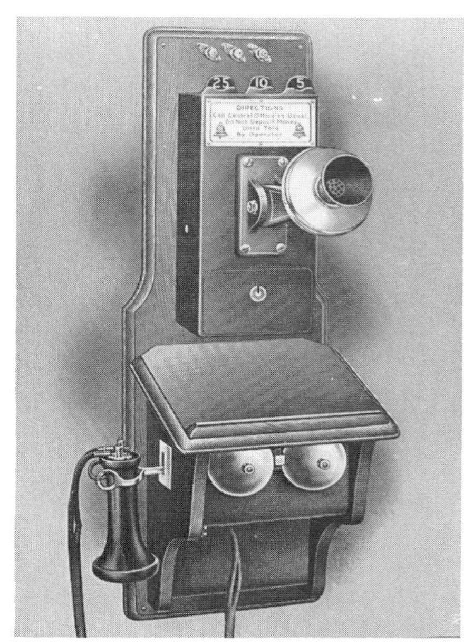

Fig. 5.17 *A wall telephone widely used in North America in the early 1900s modified by the addition of a Gray Pay Station Company coinbox*

Fig. 5.18 *Popular in the 1920s, this Shield shaped coinbox enamelled in red, white and blue made by the Gray Pay Station Company to fit on to the backboard of a wall telephone*

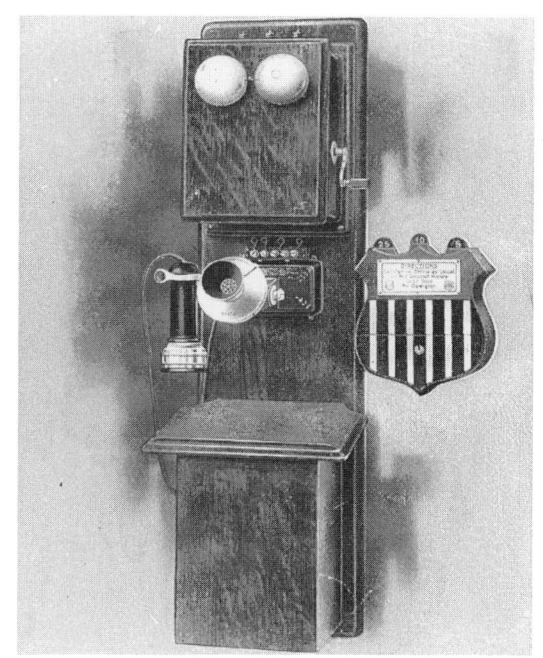

Fig. 5.19 *A version of the shield shaped coinbox made to fit alongside a wall telephone. The sound of the coin signals was conducted through a metal plate behind the telephone and the lower left hand fixing screw of the transmitter bracket to the transmitter, c. 1925*

Fig. 5.20 *A standard candlestick instrument modified by the attachment of a Gray Pay Station Company coinbox. The sound of the coin signals is conducted through the supports attaching the coinbox to the telephone, c. 1925*

Fig. 5.21 *Telephone for coal mines and other industries made by Lorenz of Berlin in the early 1900s. The bronze case is corrosion resistant and does not produce sparks if struck. At the bottom of the case are glands for telephone cables. Twin receivers inside the instrument project sound into the pivoted tubes on either side of the telephone which can be positioned near the telephone user's ears*

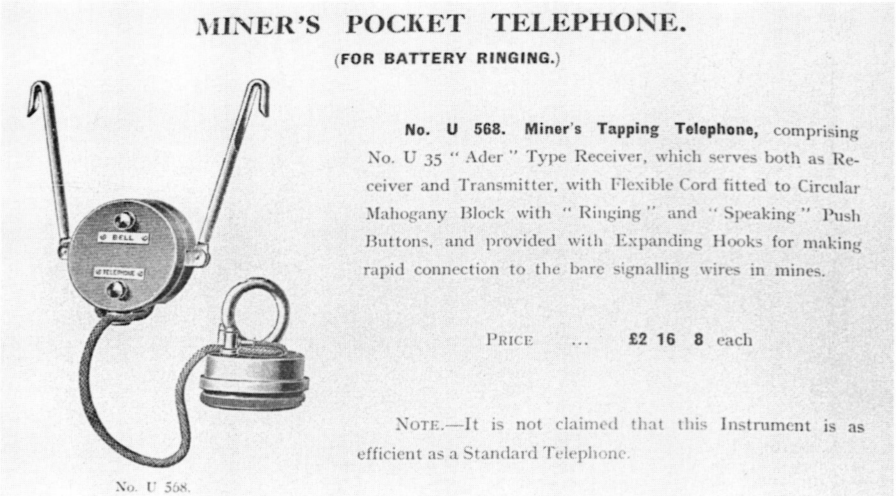

MINER'S POCKET TELEPHONE.

(FOR BATTERY RINGING.)

BELL

TELEPHONE

No. U 568. Miner's Tapping Telephone, comprising No. U 35 " Ader " Type Receiver, which serves both as Receiver and Transmitter, with Flexible Cord fitted to Circular Mahogany Block with " Ringing " and " Speaking " Push Buttons, and provided with Expanding Hooks for making rapid connection to the bare signalling wires in mines.

PRICE ... **£2 16 8** each

NOTE.—It is not claimed that this Instrument is as efficient as a Standard Telephone.

No. U 568.

Fig. 5.22 *Portable miner's telephone of the 1920s for use where there was no gas hazard. The instrument was carried in the pocket or hooked on to bare signalling wires which ran throughout the Gallery or Working. Advertised by the Sterling Telephone and Electric Company of Great Britain but probably designed in France*

Fig. 5.23 *The 'one-button' telephone was extensively used by railways as a general purpose instrument. This model dates from the late 1930s*

Fig. 5.24 *A 1930s example of 'two-button' telephone used by railways on omnibus circuits connected to a switchboard. One button called the operator, the other was used to code ring other telephones on the same circuit*

Fig. 5.25 *A selective ringing telephone of the late 1930s. Selective ringing was particularly useful on busy omnibus circuits as it minimised the disturbance caused by unwanted ringing*

Fig. 5.26 *A selective and code ringing telephone made by the Automatic Telephone Manufacturing Company of Liverpool, England in the 1920s*

Fig. 5.27 *A superimposed telephone of the early 1920s manufactured by the Automatic Telephone Manufacturing Company of Liverpool, England, and marketed under the trade name 'Phantophone'*

Fig. 5.28 *Phonopore telephone with two Ader receivers for conversation and a Collier-Marr receiver, at the top of the telephone, to reproduce the calling signal. This was used in Britain by the North Eastern Railway in the early 1900s. The Collier-Marr receiver has suffered the loss of the small horn which magnified the sound of the calling signal*

Fig. 5.29 *A battery of railway telephones in the Taunton East signal box of British Rail photographed in November 1983 shortly before the box was closed*

To overcome this danger, L.M. Ericsson produced a specially modified telephone. The instrument itself was fitted high up on the wall, well beyond the reach of a person standing on the ground. Rubber tubes, hanging from the transmitter and the receiver, enabled a person standing on the ground to speak and to listen. The magneto-generator was turned by a pulley driven by a rope from the lower level where there was a second pulley turned by a cranked handle. Later, L.M. Ericsson designed a handset version of this instrument in which both tubes terminated in what was superficially a conventional handset, although there were no electrical components within it. A telephone for use in similar circumstances was also made by the Bell Telephone Manufacturing Company of Antwerp.

Today, there is still a need for telephone communication between power stations but, instead of using separate telephone wires, it is common practice to carry calls on the same wires that carry power, by using different frequencies. When this is done, standard telephone instruments can be used without danger. This is because safety equipment is incorporated in the apparatus which combines and separates the frequencies used by the telephone and those used for power.

People in many trades and professions benefited from the invention of the telephone but there were few whose working conditions and safety improved to such an extent as divers. Before the second half of the nineteenth century, divers needed to be men of immense courage. Diving suits frequently leaked and divers

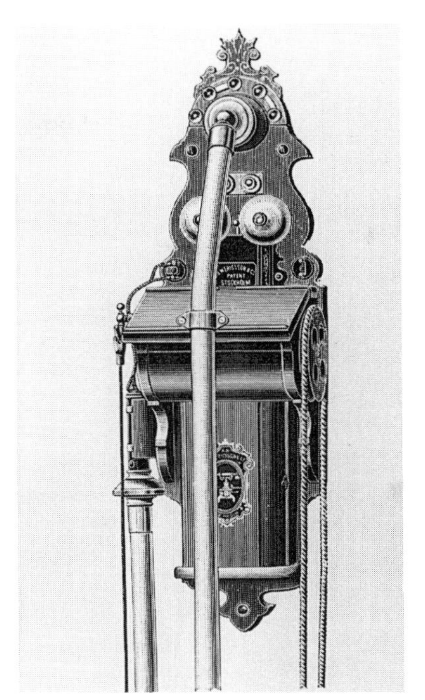

Fig. 5.30 *A telephone that could be safely used where there was a risk of contact with high voltage. Made by L.M. Ericsson in the early 1900s*

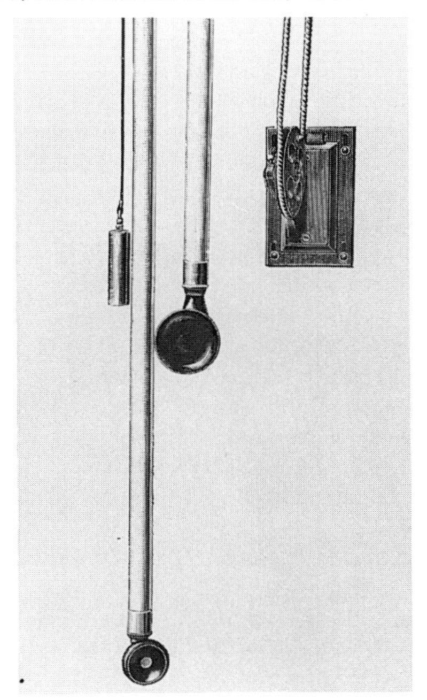

were dependent on air pumped down to them, not only to breathe, but to supply oxygen to their tiny oil lamps which, in murky waters, were the only means by which they could see. To communicate with the surface they tugged on a rope using a system of pre-arranged signals and when two divers worked together they had to use hand signals.

A vast improvement in divers' conditions was brought about by the application of the telephone but a number of factors impaired its efficiency. The inside of a diver's helmet was a far from ideal place in which to install a telephone. For one thing, it was generally wet, and salt water is an extremely good conductor of electricity. This wetness was partly caused by seepage and partly by condensation. There was also a third cause which was the operation by the diver

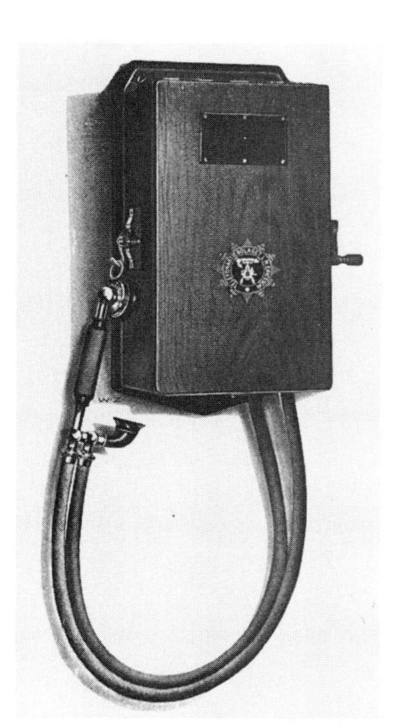

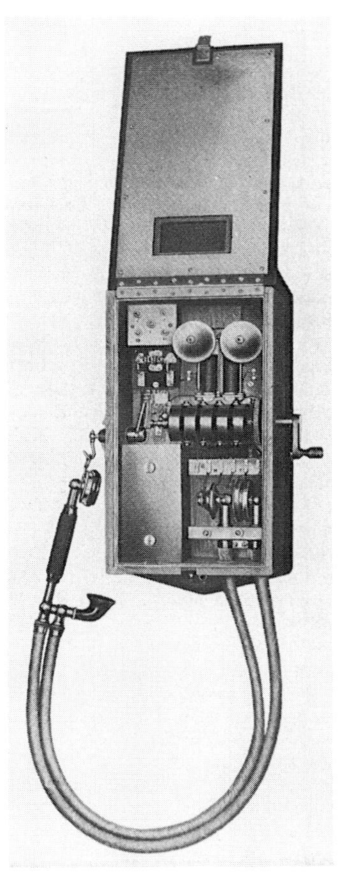

Fig. 5.31 *Telephone with protection from high voltage made by L.M. Ericsson in the 1920s. The handset, although apparently conventional, contains no electrical components. It is linked to a transmitter and receiver within the wooden case by rubber speaking tubes*

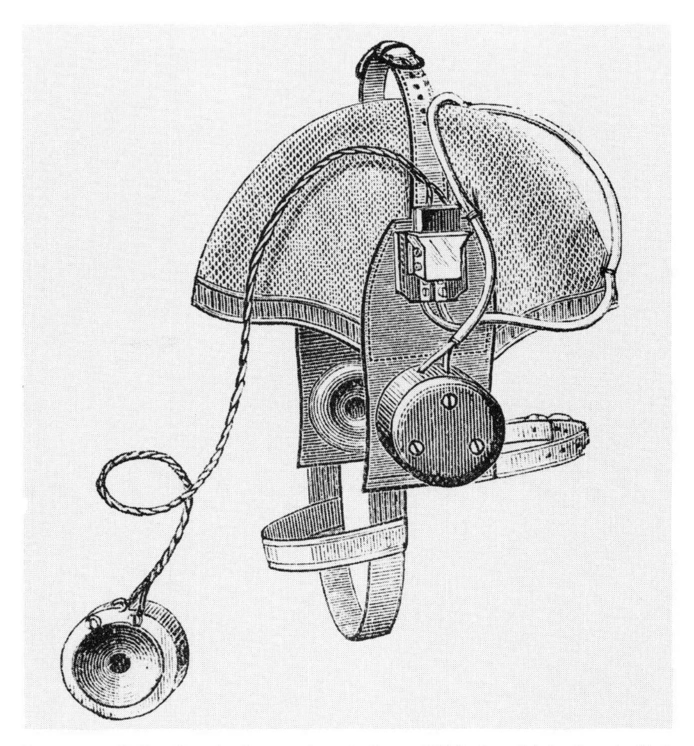

Fig. 5.32 *One type of diver's telephone, dated about 1900, in which closely fitting earphones were used to keep out background noise. The transmitter was attached to the inside of the helmet at one side of the face glass*

of the 'spitcock'.[19] This was a small tap situated in front of the helmet near the diver's mouth. By operating this tap the diver could take into his mouth a small quantity of water which he could then direct at the face glass, or window, to clear it of condensation. The telephone not only had to work in the wet, it also had to work in compressed air. This made it harder for the moving parts of the transmitter and receiver to vibrate, which in turn reduced the volume of the transmitted speech. Lower volume combined with the background noise caused by compressed air entering the helmet, and the bubbles leaving it, sometimes made it difficult for the diver to hear. He sometimes overcame this problem by temporarily turning his air supply off – but his messages would have to be brief and this was not, perhaps, the ideal circumstance in which to develop the art of good conversation.

In 1907, the Cunard Steamship Company commissioned the National Telephone Company in Britain to install central battery telephone systems in two of its trans-Atlantic liners, the 'Mauretania' and the ill-fated 'Lusitania'. Faced with the problem of keeping the transmitter as still as possible when the ship

Fig. 5.33 *Telephone used for speaking to divers. This instrument, of about 1900, could be used with two divers enabling them to speak to each other as well as to the surface*

might be rolling or pitching, the telephone company adopted a hand-held instrument which, when not in use, was retained by a clip on the wall. Another clip prevented the receiver being shaken from its rest in rough seas. The only passengers to be provided with telephones were those in the 'regal' and 'first class' state rooms. The instruments were either gilded or silvered to blend with the state room's individual decor. The 'Lusitania' is now best remembered because, on 17 January 1915, it was sunk by a torpedo from a U boat and this incident was one of the factors which led to the USA entering the First World War.

By the Edwardian era, most people knew that germs could produce diseases and many were concerned – perhaps over-concerned – about infection. Some people were reluctant to use telephones because they were afraid of contracting a disease from an instrument that had been used by an infected person.

Several inventors conceived devices intended to eliminate this risk. One depended on the disinfectant properties of the gas, ozone. This is a form of

Fig. 5.34 *One of the hand-held telephones used on the Cunard liners, 'Lusitania' and 'Mauretania', c. 1907*

oxygen which can be produced from air by electrical sparks. The inventor proposed to use the voltage from the telephone's magneto for this purpose and to blow the gas into the telephone's mouthpiece and earpiece with a tiny bellows which was also worked when the magneto handle was turned. As far as is known, this telephone was never made.[20]

A sanitary idea that was not only made, but also used commercially for many years, was the glass mouthpiece. It was never a standard part of a particular telephone, but was made by several manufacturers as an accessory that subscribers could buy to replace the standard mouthpiece. There were versions for both fixed transmitter and handset telephones. Manufacturers recommended that 'Mouthpieces should be boiled for one minute as often as necessary.'

Perhaps the strangest, and least practical of all the sanitary ideas, was a telephone in which the mouthpiece and earpiece were sterilised by the heat from a gas flame.[21] Whenever the receiver was lifted or replaced, the gas was turned on and ignited by a hammer striking and exploding a paper cap. Heat from the burning gas was applied to the mouthpiece and earpiece and also caused mercury

"STERLING" SANITARY DETACHABLE GLASS MOUTHPIECE.

SANCTIONED FOR USE ON G.P.O. WALL OR TABLE TELEPHONES.

EASILY REMOVED AND CLEANED. EASILY REPLACED.

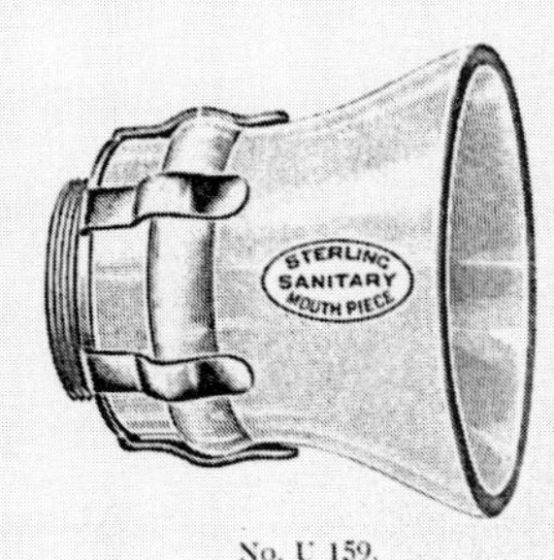

No. U 159.

No. U 159.—Glass Mouthpiece, suitable for Sterling Wall or Table Telephones Nos. U 55, U 206, U 250, U 320, U 325, U 345, U 347, U 348, U 352, U 353, U 702, U 706, and U 715.

PRICE **2/-** each.

By applying a slight Pressure to one side of the Mouthpiece the Glass can be removed for Cleaning and replaced in the Metal Clips without unscrewing the Holder.

Shewing No. U 159 fitted to No. U 715 C.B. Table Telephone.

No. U 160.—Glass Mouthpiece suitable for Sterling Hand Combinations Nos. U 46, U 95, U 105, U 132, U 165, U 170, U 175, U 180, U 185, U 190, U 292, U 305, U 306, U 335, U 345, U 385, U 412, U 426, U 427, U 440, U 442, U 445, U 447, U 452, U 453, U 454, U 486, U 491, U 505, U 510, and U 716.

PRICE **2/6** each.

The Mouthpiece can be removed for cleaning by pulling it out of the Transmitter Case, and replaced by pushing it home again.

Showing No. U 160 fitted to No. U 175 Hand Combination.

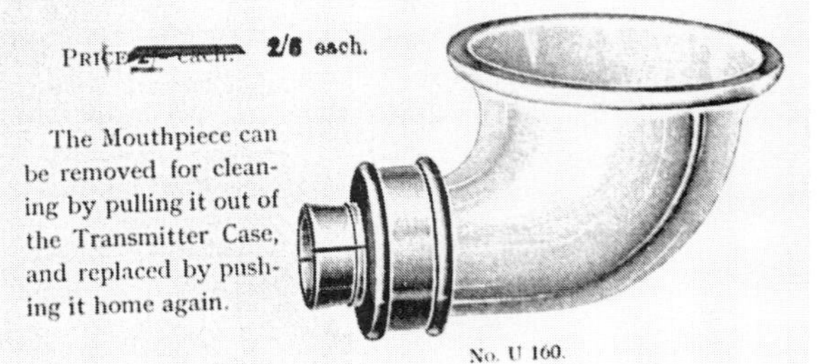

No. U 160.

Fig. 5.35 *A page from the Sterling Telephone Company's catalogue of the 1920s*

Fig. 5.36 *By the early 1900s telephone manufacturing companies were developing multinationally. This L.M. Ericsson advertisement of the period shows that the firm had works in Sweden, Great Britain, Russia and the USA.*

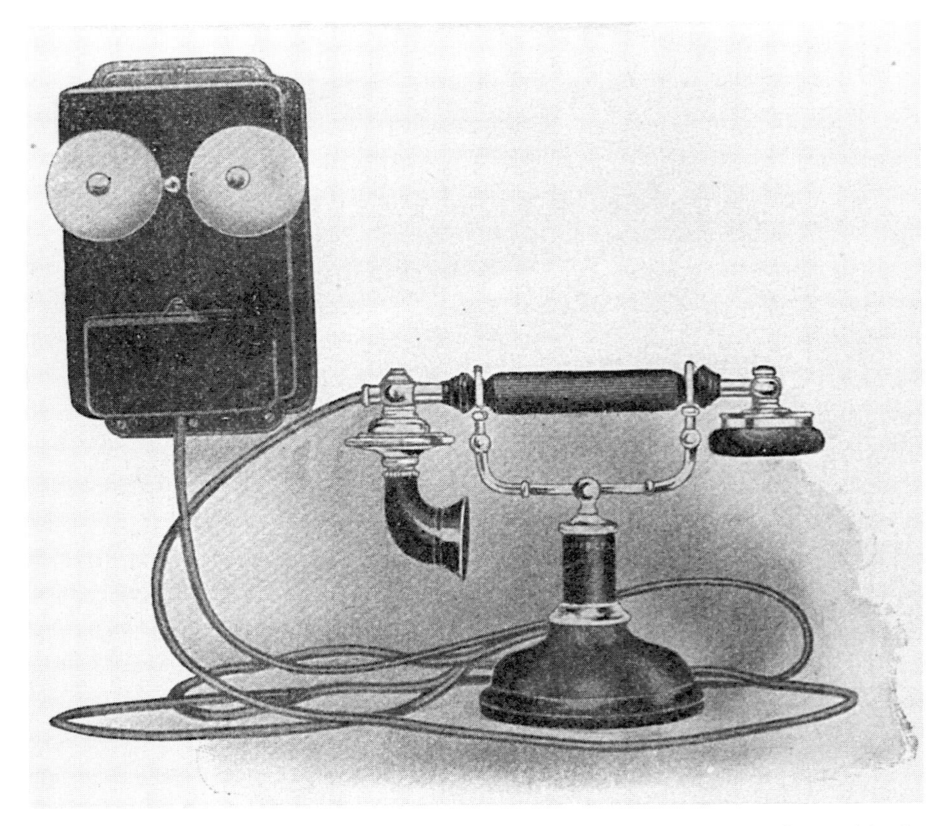

Fig. 5.37 *A very rare American handset telephone of the early 1900s manufactured in the Buffalo, New York, workshops of the Ericsson Telephone Manufacturing Company in an unsuccessful attempt to popularise the handset in the USA. This company was a subsidiary of L.M. Ericsson of Sweden*

in a cylinder to expand and move a piston. After a brief interval, the movement of this piston turned off the gas. As far as is known, telephones using this idea were never manufactured, which was probably just as well, considering the dire consequences which might have arisen from a malfunction of the equipment. In those days when beards, side-whiskers and moustaches were commonplace, one can imagine a newly-beardless subscriber demanding that the telephone company immediately remove their flaming telephone.

By the early 1900s the leading telephone manufacturing companies were evolving into important multi-national concerns. Several firms had foreign subsidiaries or associates making telephones designed by the parent organisation. It therefore became progressively more difficult and less relevant to classify telephones by 'country of origin'.

Chapter 6

New telephones for new exchanges

In the early days, a battery in or near each telephone provided current for the carbon transmitter. This arrangement gave good results as long as the battery was well maintained. However, telephone administrations found that battery maintenance was expensive. A way of saving this cost was devised quite early in the history of the telephone and even used commercially.[22] This was to have a large battery at the exchange which supplied power to the individual telephones over the same wires that carried speech. Unfortunately, in those days, this could only be done at the expense of reducing telephone efficiency. This meant that speech was faint, particularly on longer lines. Completely successful ways of supplying power from exchange batteries were eventually devised by H.V. Hayes and J.S. Stone of the USA and the first exchanges to use this technology came into use in the 1890s. The system was called 'Common Battery' in the USA and 'Central Battery' in Britain. Both names were abbreviated to 'CB'.

Compared with older systems, CB was less tolerant of the incidental variations in electrical resistance of telephone transmitters which occurred when they were moved. This was particularly likely to occur when transmitters were mounted in handsets. A number of European manufacturers contended that their handsets could be used with CB exchanges. However, telephone administrations were more critical. In most cases, where, as in Europe, handsets had been used, they were removed when CB was introduced. This was done with reluctance because handsets were popular with telephone users. For the same reason, handsets used with local battery exchanges were allowed to stay. In the USA handsets had never been adopted for general use and the introduction of CB caused little or no change in the superficial appearance of telephones. This also applied to Canada which closely followed the practice of the USA.

In the years following 1900, the introduction of CB brought about two, quite separate, lines of telephone evolution – fixed transmitter telephones and handset telephones. Fixed transmitter telephones had the advantage of versatility because they could be designed to work with either CB or local battery exchanges. However, in spite of this advantage, they tended to become stereotyped in both Europe and North America and there were relatively few

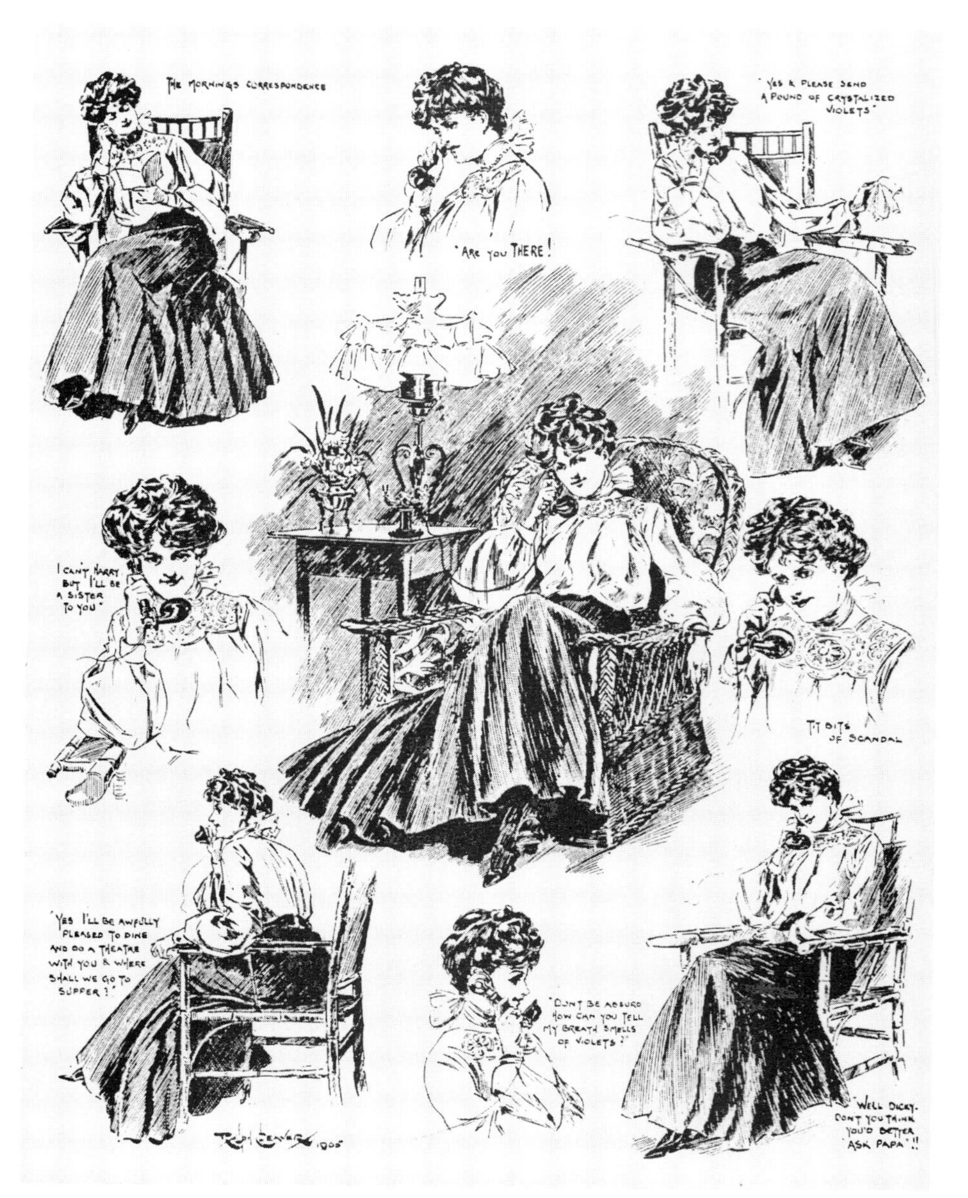

Fig. 6.1 *In Britain, Edwardian subscribers appreciated the ease and comfort with which handsets could be used*

Fig. 6.2 *A British table telephone of 1900 standing on a wall-mounted battery box which also functioned as a telephone shelf. The handset is a British-made version of an L.M. Ericsson design. The hook on the right is to accommodate an additional earpiece if required*

Fig. 6.3 *The telephones of the Swedish manufacturer, Ericsson, were always noted for quality but when called upon the firm's craftsmen could excel even their own high standards. This instrument specially made in gold, ivory and engraved steel was delivered to the Kremlin in 1903 and used by Tsar Nicholas II*

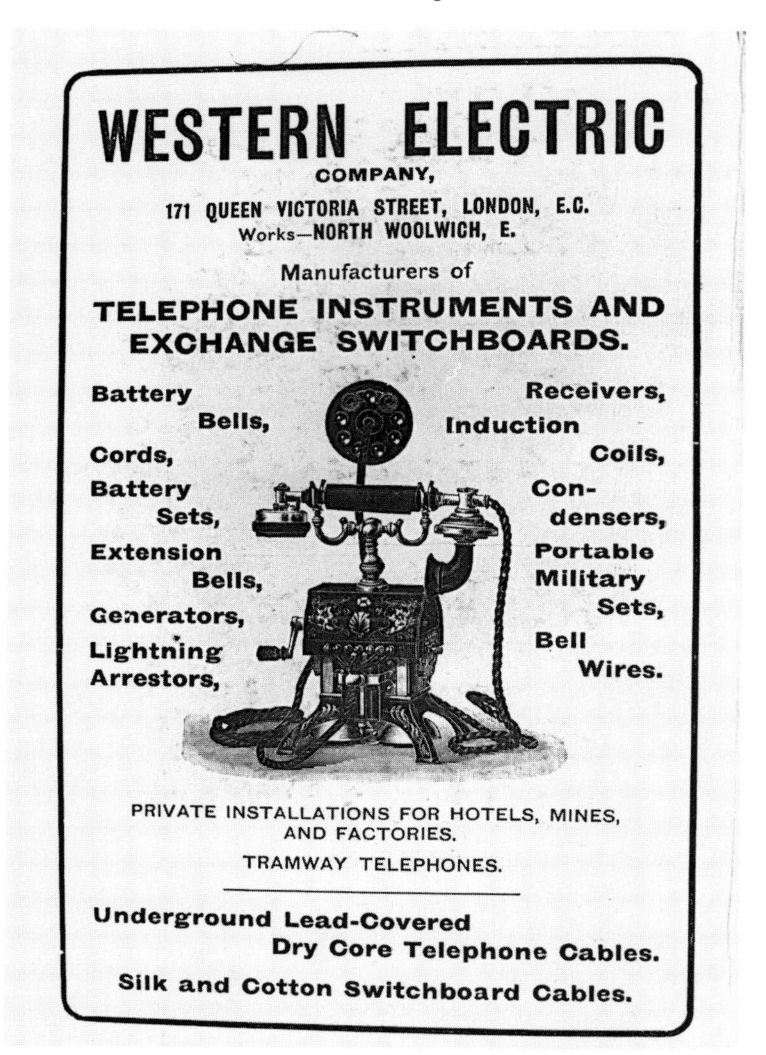

Fig. 6.4 *1906 advertisement*

basic types. By contrast, in Europe at least, handset telephones continued to be manufactured in diverse profusion.

In Europe, telephone subscribers connected to CB exchanges missed the convenience of handsets. In particular, people who were taller or shorter than average found fixed transmitter telephones awkward to use and, most unpopular of all, were telephones that were wall-mounted. There were several inventions that were intended to make fixed transmitter telephones more acceptable by

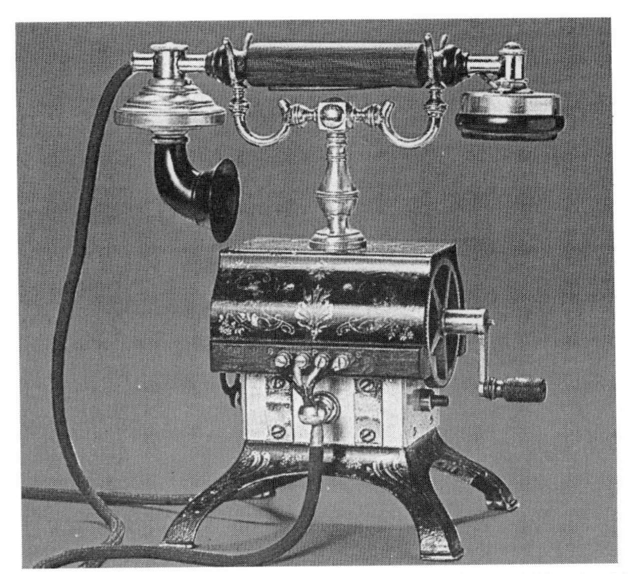

Fig. 6.5 *Telephone of 1902 used in Portugal and manufactured by the Western Electric Company of London. Even well preserved telephones are sometimes incomplete. This one lacks a cog from the gear through which the handle drove the magneto*

Fig. 6.6 *This walnut table telephone of 1900, nicknamed 'Sabrina', was used in Britain by the National Telephone Company. There were two versions, one with fixed transmitter and one, as shown, with handset. Both were made by the Western Electric Company in England, and by the Bell Manufacturing Company in Belgium*

Fig. 6.7 *A 20 cent stamp of Bophuthatswana illustrating the fixed transmitter version of the telephone nicknamed 'Sabrina'. This instrument was made for the southern African market by the Swiss firm of Hasler, under licence from the Bell Telephone Manufacturing Company of Antwerp, Belgium*

providing means of raising or lowering the transmitter. However, in order to maintain the transmitter's efficiency, this had to be done without causing it to tilt.

In Germany, the telephone manufacturer Mix and Genest produced a telephone in which the transmitter was supported by two separate arms mounted one above the other. Both arms were hinged where they joined the front of the telephone, and also where they joined the transmitter, so that the entire assembly formed a parallelogram. A spring between the arms pulled the transmitter upward in order to counter the downward pull of gravity. By this means the transmitter could be easily raised or lowered without tilting. In Britain, a similar adjustable bracket was manufactured by the Sterling Telephone Company.

There was quite a different approach to the problem by British Insulated and Helsby Cables Ltd. It was called the 'radial arm' and consisted of an arm which could be rotated like the hand of a clock. The transmitter was mounted at the end of this arm and by turning the assembly the height of the transmitter could be varied by six inches.

Fig. 6.8 *Two British telephones of 1900 used by the National Telephone Company. The handset telephone was used with local battery exchanges and the fixed transmitter version was used with central battery exchanges*

Fig. 6.9 *A telephone of 1900 used in Germany. The two magneto handles enabled it to be used from either side. This was a convenience when two people sat on opposite sides of the same desk*

Over the years, telephone administrations have charged for their services in a number of ways. The simplest method was the 'flat rate' or 'inclusive charge'. This included the cost of all calls made within the local area. The inclusive charge was attractive to telephone administrations, particularly in their early years, because it enabled income to be predicted and made development planning easier. But the inclusive charge was not universally popular with subscribers. In Britain, in particular, they pressed for charges to be based on the number of calls they made. This charging method was called the 'measured rate'. What eventually emerged, in most countries of the world, was a compromise between these conflicting points of view. Subscribers paid a considerably reduced rental plus a charge for every successful call they made. However, there are many local variations of this system.

The introduction of the measured rate made it necessary for telephone administrations to keep a record of all calls. Initially this was done manually but later, meters were introduced. The location of the meter was a further subject for consideration. Given the choice, many subscribers would have liked it to be at, or

Fig. 6.10 *A French table telephone of 1910. The handset was a standard type fitted to both wall and table instruments. It had a loop with which it was hung up when used with other types of telephone*

Fig. 6.11 *Telephone used in Holland and sometimes called the 'cotton reel' telephone. Manufactured by L.M. Ericsson of Sweden, early 20th century*

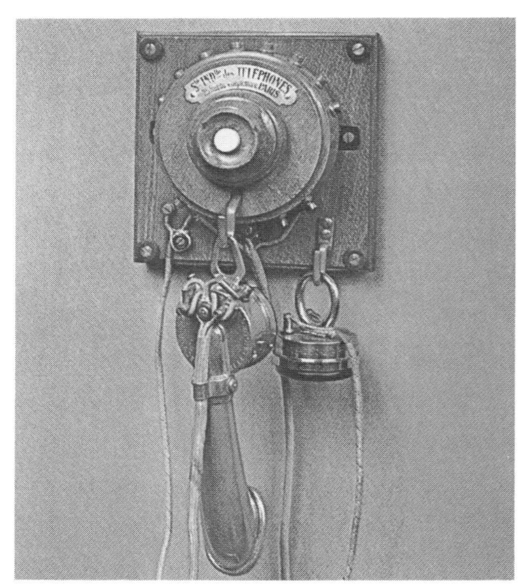

Fig. 6.12 *French telephone manufactured by S.I.T. in 1903. The receiver and transmitter were mounted back to back and both held to the ear when in use. In this position transmitter movement was minimised. The horn, which also acts as a speaking tube, conveys sound to the transmitter from the mouth. This type of handset was sometimes called a 'monophone'.*

near, the telephone instrument as this would enable them to monitor the cost of calls. Telephone administrations, on the other hand, benefited from having all the meters together in the exchange because this arrangement facilitated meter reading. In the event, the location of the meter was determined by practical considerations. Because calls were connected at the exchange, the exchange was the best place for the meter. Any other arrangement involved transferring information about the progress of calls from the exchange to the distant meter. This was bound to introduce added complexity and increase the risk of error. Even so, a number of inventors have tried to find a practical solution to this problem with varying degrees of success.

In 1894, W.T. Gentry of the Southern Bell Telephone and Telegraph Company of Atlanta, Georgia, devised a simple telephone call meter based on the cyclometer. This meter had a plunger which was operated by the subscriber. It was manufactured by the Gray Telephone Pay Station Company and used with several types of telephone. A similar meter was used in Germany in the early part of the 20th century.

Several companies not otherwise connected with the telephone business, marketed counters which could be placed near the telephone. There was, of course, no certainty that the number of calls recorded by the subscriber equalled

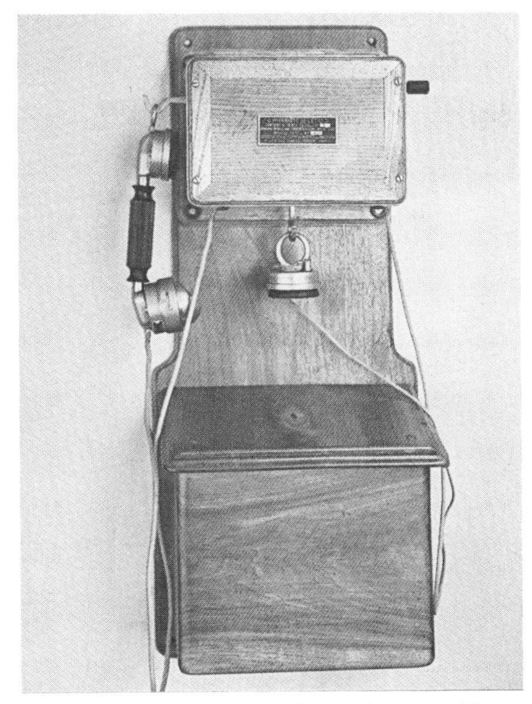

Fig. 6.13 *A wall telephone of 1910 made in France for use with a magneto type manual telephone exchange*

Fig. 6.14 *German table telephone of 1900 made by Mix and Genest*

Fig. 6.15 *A spectacular German telephone of 1900 made by Mix and Genest*

the number recorded at the exchange. The solution to that problem came much later after further technical advances had been made.

A further complication was that, as today, trunk calls were timed and charged for according to their duration. This added a further difficulty for people who wished to monitor the cost of their calls. An early mention of a subscriber's timing device was contained in an 1895 report written by a Victorian telephone engineer who travelled extensively in Europe. Referring to the telephone instruments of Würtemberg – later to become part of Germany – he wrote: 'These comprise magnetos, Berliner transmitters, and spoon-shaped double-pole receivers. Some are fitted with sand-glasses to enable subscribers to time their trunk conversations.'

On 10 September 1879, exactly three and half years after Bell made the first successful telephone call, three Americans: M.D. Connolly, T.A. Connolly and T.J. McTighe, applied for a patent for an automatic telephone exchange.[24] The patent also contained a description of a rudimentary telephone dial. Although it was never used commercially, their idea was basically sound and a working system was exhibited at l'éxposition d'électricité in Paris in 1881.[25]

Fig. 6.16 *This telephone, made by Rudolph Kruger of Berlin in the 1890s, had a moveable tray which operated the gravity switch when the handset was rested on it. This tray was responsible for the instrument's nickname—the 'ashtray telephone'*

For an idea as complex as an automatic telephone exchange to be thought of so soon after the invention of the telephone is remarkable enough, it is even more remarkable that several other inventors were thinking on similar lines.

In America, George Westinghouse applied for a patent only 31 days after Connolly, Connolly and McTighe.[26] However, the Westinghouse system was not a fully automatic exchange in which subscribers controlled the equipment, but what was later called a 'semi-automatic system', in which equipment was remotely controlled by a telephone operator from a distant exchange. The purpose of the system was to save the expense of a telephone operator in small uneconomical exchanges. It was also claimed that the semi-automatic system avoided the need for subscribers to learn to operate automatic equipment.

In Britain, a semi-automatic switching system was patented by Dane Sinclair on 7 July 1883.[27] It was first used commercially at Coatbridge in Scotland and later at several other places.

In Sweden, semi-automatic systems also date from 1883. In that year, Lars Magnus Ericsson designed equipment for up to ten subscribers. This continued to be used in several places until well into the 20th century.

The next major advance came in 1889 when an American, Almon Brown Strowger, devised a new type of selector for automatically connecting telephone

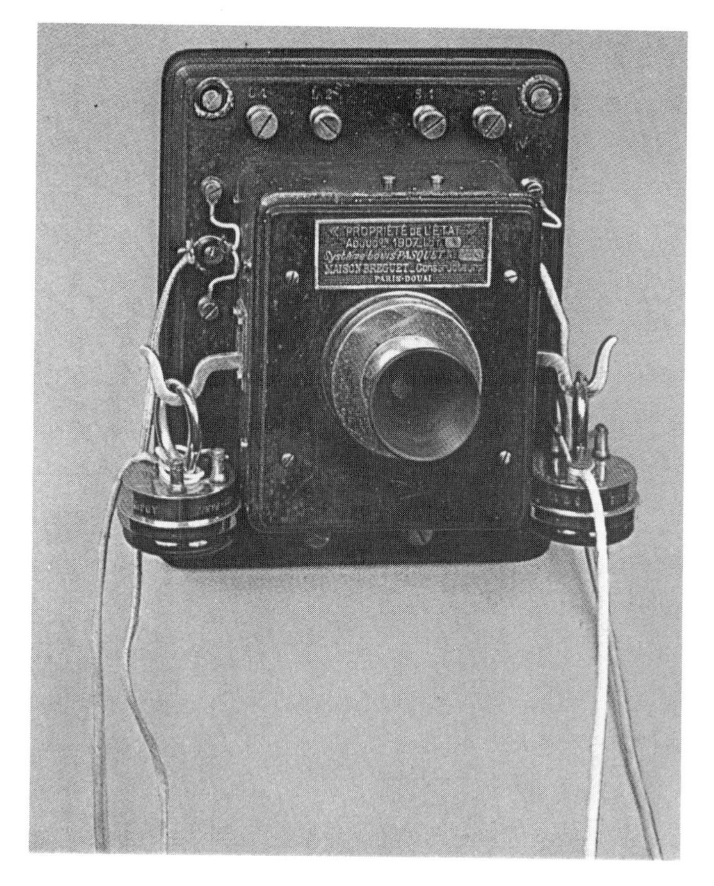

Fig. 6.17 *French telephones with twin 'bull-ring' receivers made by Pasquet in 1902*

Fig. 6.18 *German table telephone of 1900 with twin receivers and a transmitter mounted on an adjustable arm*

calls.[28] In 1892 he installed the world's first fully automatic public telephone exchange at La Porte, Indiana, USA. It commenced working on 3 November but, owing to manufacturing difficulties, Strowger's new mechanism was not ready and more rudimentary equipment had to be used. The new mechanism was first used in June 1895, also at La Porte.

At this point in its development, the Strowger system was suitable only for the smallest exchanges. This was changed during the latter half of the 1890s by three members of the Strowger team, A.E. Keith, J. Erickson and C.J. Erickson who evolved ways of interconnecting and combining the Strowger equipment so as to form telephone exchanges of unlimited size. With this development the future of automatic exchanges was assured.[29]

In Europe, the first fully automatic exchange was opened in Amsterdam in June 1898.

Other successful telephone exchange systems were developed later but, during the first half of the 20th century, the Strowger system was the most successful and widely used.

Fig. 6.19 *In the USA, the Bell telephone receiver became such a familiar object that there was no need to explain what the flower in this 1902 advertisement of the Twin City Telephone Company was meant to represent. This company operated in Minneapolis and St. Paul, Minnesota*

Fig. 6.20 *As this postcard of the early 1900s shows, the inconvenience of the fixed transmitter telephone was soon forgotten when the right person was at the other end of the line*

Fig. 6.21 *A British candlestick telephone of 1905 for use with central battery exchanges. The transmitter is a solid-back type designed by A.C. White of the USA and made by the Western Electric Company. The bright metal parts were plated, generally with nickel but occasionally with silver. The bulbous section in the middle of the stem was responsible for the instrument's nickname – the 'golf ball' telephone*

Subscribers connected to automatic exchanges needed a means of signalling the number they wanted to the exchange. The Connolly brothers and McTighe adapted a telegraph instrument to form a rudimentary telephone dial but it was only required to send one digit.

Strowger at first employed press buttons and in his first patent a telephone with four press buttons is shown.[28] The first three buttons were used to transmit the hundred, ten and unit digits by depressing each button the appropriate number of times. The fourth button was used to release the exchange equipment at the end of the call. Later, Strowger and his associates experimented with dials and the earliest version of the dial, as we know it today, in which digits are signalled in sequence, emerged in 1896.

Other devices to signal the wanted number to the exchange were used by various companies but, without question, the dial was the most successful. It was not only used with the Strowger system, it was adopted and used with most other

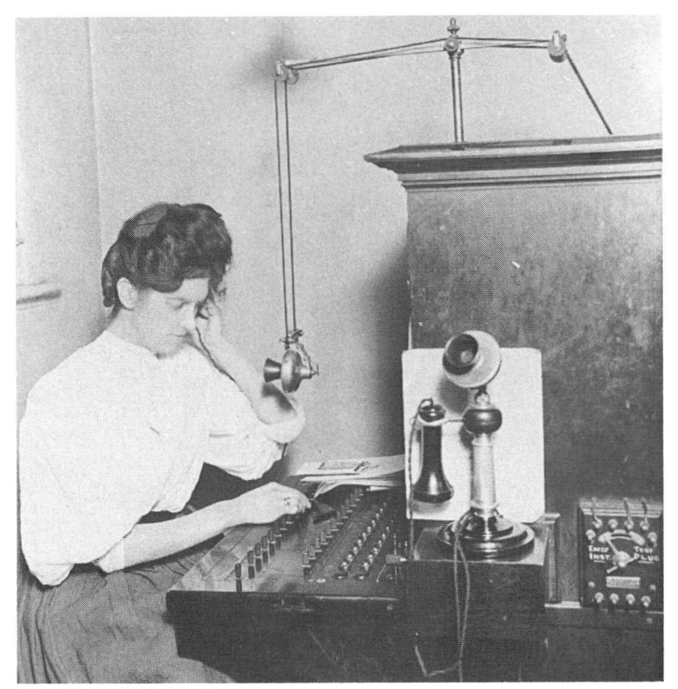

Fig. 6.22 *In the early 1900s, the 'leg o'mutton' sleeves worn by this swtichboard operator were slightly outmoded. The 'golf ball candlestick' telephone, by contrast, was then the very latest type*

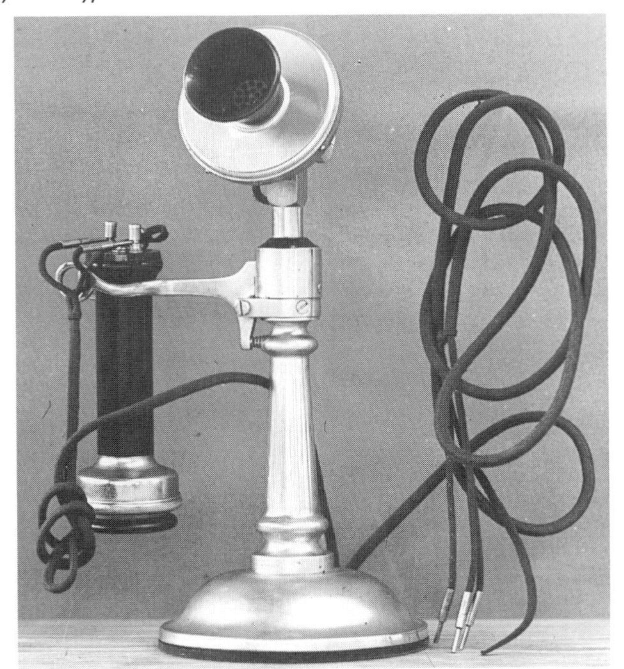

Fig. 6.23 *A central battery desk telephone of 1907 manufactured by the Western Electric Company*

Fig. 6.24 *Illustration used for the 1901 song 'Hello, Central, Give me Heaven' by Charles K. Harris. The picture also shows in an acute form, the problem of the fixed transmitter telephone*

telephone exchange systems. Indeed, there was no real challenge to its supremacy until press button signalling became practical with the development of electronics.

An incidental result of the success of the dial was that the words 'dial telephone' were used by people in North America as a colloquialism to signify the

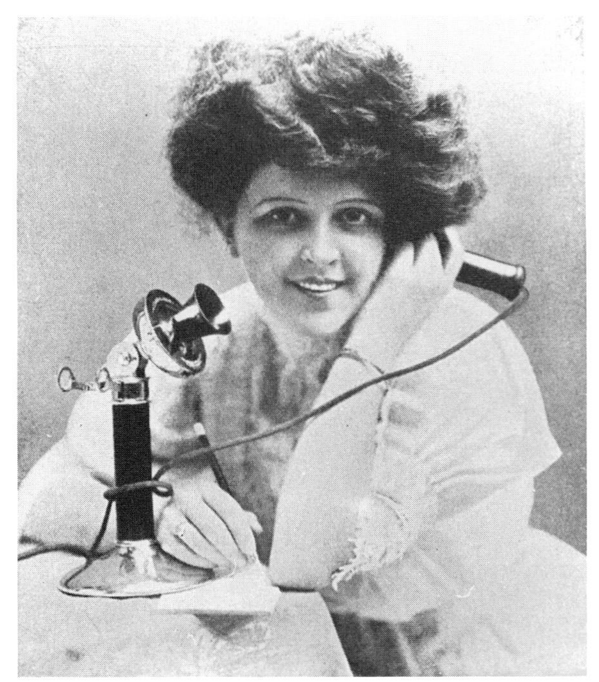

Fig. 6.25 *Picture from an almanac of 1910 published in Birmingham, England*

Fig. 6.26 *Early 20th century illustration produced by the Danish telephone administration 'Kjøbenhavns Telefon Aktieselskab' (KTAS) depicting the advantage of the adjustable height transmitter*

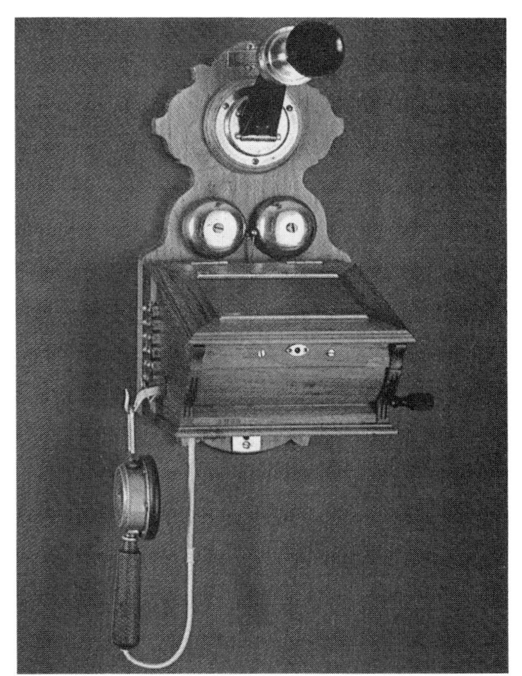

Fig. 6.27 *Telephone made by Mix and Genest introduced in 1904. The transmitter is on a special bracket which enables it to be raised or lowered without tilting*

Fig. 6.28 *Radial arm telephone which allowed the height of the transmitter to be varied by six inches*

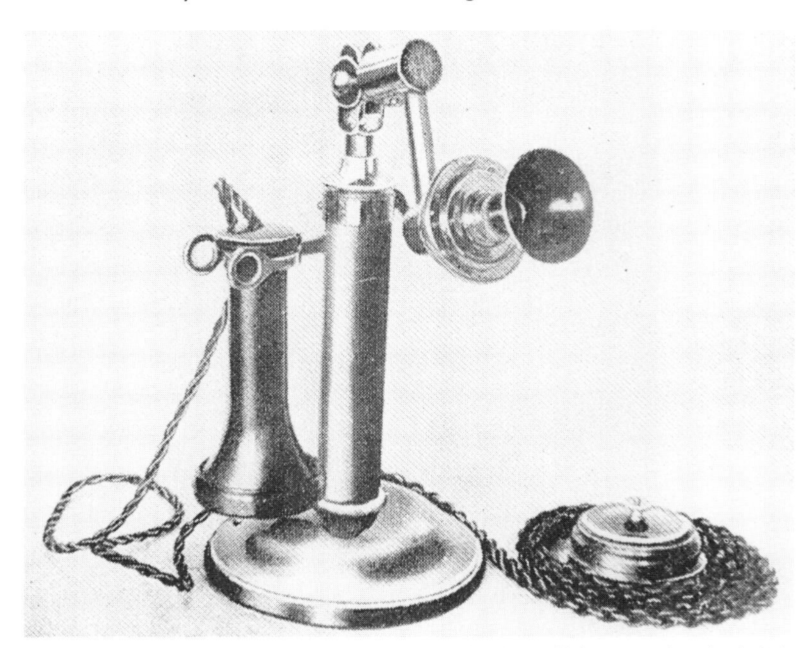

Fig. 6.29 *An unusual candlestick telephone of 1910 with radial arm to alter the height of the transmitter*

Fig. 6.30 *A metal-cased telephone with radial arm bent into a crank to minimise the amount by which the transmitter protruded from the wall*

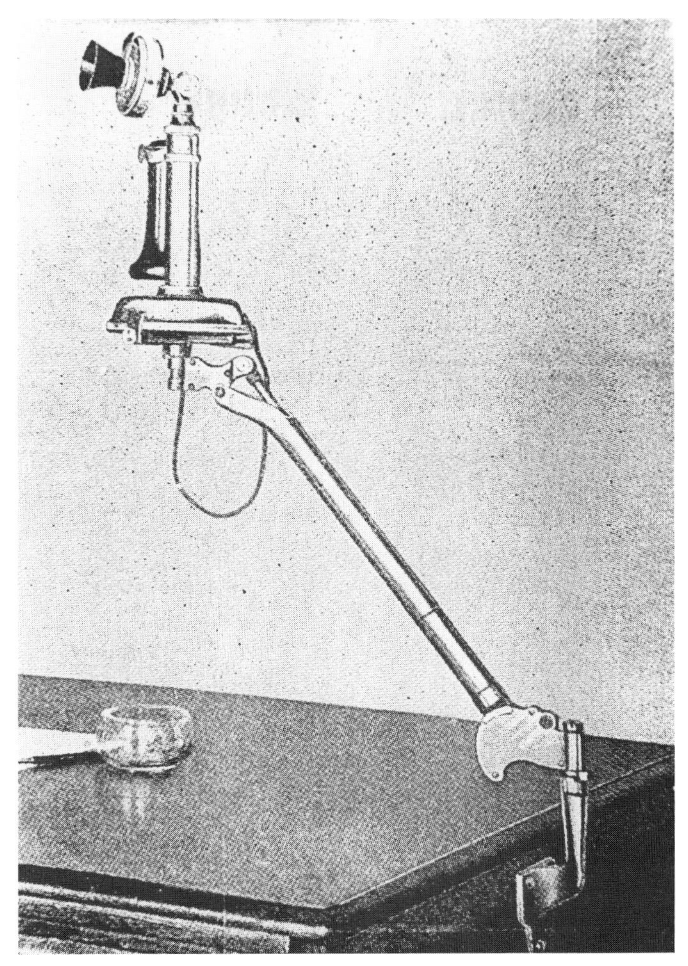

Fig. 6.31 *A comprehensive solution to the problem of making the fixed transmitter telephone more convenient to use. This bracket of 1910 called the 'Equipoise Telephone Arm', supported the entire telephone*

automatic telephone service. This form of expression is seldom, if ever, heard on the opposite side of the Atlantic.

Until the Strowger dial achieved its supremacy there were other devices with which subscribers could signal the number they wanted. Some were different types of dial, while others were not dials at all. Probably the most unconventional method of signalling was used with the Lorimer system developed in Canada and the USA by three brothers, George, Holt and Egbert Lorimer. The system was patented in 1900.

Fig. 6.32 *An inventor's idea of 1910. A telephone that could be used with the hands free. The receiver remained attached to the bracket and a clip held the receiver rest down when the telephone was not in use. As far as is known it was never manufactured*[23]

Lorimer exchanges were manufactured by the Canadian Machine Telephone Company of Toronto and also by the American Machine Telephone Company Ltd. They were installed in several places in Canada and the USA and also in Hereford, England; Lyons, France; and Rome. In the Lorimer system, the number the subscriber wanted was signalled to the exchange with a 'signal transmitter'. This had a number of levers in a row. Associated with each lever was a small hole or 'window' through which a numeral could be seen. By moving the lever the numeral in the window could be altered. To make a call the subscriber adjusted the row of levers so that the wanted number appeared in the windows. He then turned a crank on the side of the telephone through one revolution. Instead of a crank some telephones were fitted with a handle which the subscriber pulled down. In either case, this action tensioned a spring which powered the signal transmitter and also sent a signal to the exchange to indicate that the subscriber wished to make a call. It was only at this point that the subscriber

Fig. 6.33 *Telephone call meter devised by W.T. Gentry of the Southern Bell Telephone Company of Atlanta, Georgia, fitted to an American 1920s desk telephone*

Fig. 6.34 *Gentry's telephone call meter, manufactured by the Gray Telephone Pay Station Company and fitted to several types of telephone including this early 1900s American wall instrument*

Fig. 6.35

Telephone used on party lines made by TELEGRAPHEN WERKSTATT (the factory of the Post & Telegraph Administration), Stuttgart, Germany in the 1890s. The meter, under the transmitter, recorded the number of calls and the indicator on the bell box showed if the line was engaged

Check Your CALLS with a "Checko"

Place on record every call you make and check your telephone bills.

These ingenious contrivances enable every 'phone user whether private or commercial to arrest overcharges on telephone bills.

"Checko" is what the World has been waiting for. They last a lifetime and cost but a few shillings.

Business men will appreciate "TEKO No. 1." Every call registered is indicated by the ringing of a bell, so that others can hear the call registered.

"TEKO No. 1." Register, finished in Black. Bell notifies each call registered. Automatically resets after 9,999. Sent securely packed post free. **10/6**

Adaptable for hanging on wall or for desk.

"CHECKO DE LUXE." 'Phone Call Register. Finished in nickel and mounted on Black stand. Artistic and effective. Registers silently to 10,000 calls. Adjustable reset mechanism marking back to zero at any time. Sent securely packed, post free **12/6**

"CHECKO ADAPTABLE." Small nickel-plated Telephone Call Register in cylinder form. Arranged for attaching to desk, wall or elsewhere. Registers up to 10,000 calls and automatically returns to zero. Sent securely packed, post free **7/6**

The newspaper campaign of protest and the unvoiced complaint of every telephone user are met by means of this instrument. If you instal one you can contest whatever overcharge is made for your telephone—you have direct evidence of all calls made. Post all orders, enclosing remittance and stating model required, immediately to—

MESSRS. F. & B.
(Dept. D.M.), 135, High Holborn, London, W.C.2. A. K. Co., Ld

Fig. 6.36

Newspaper advertisement of the 1920s

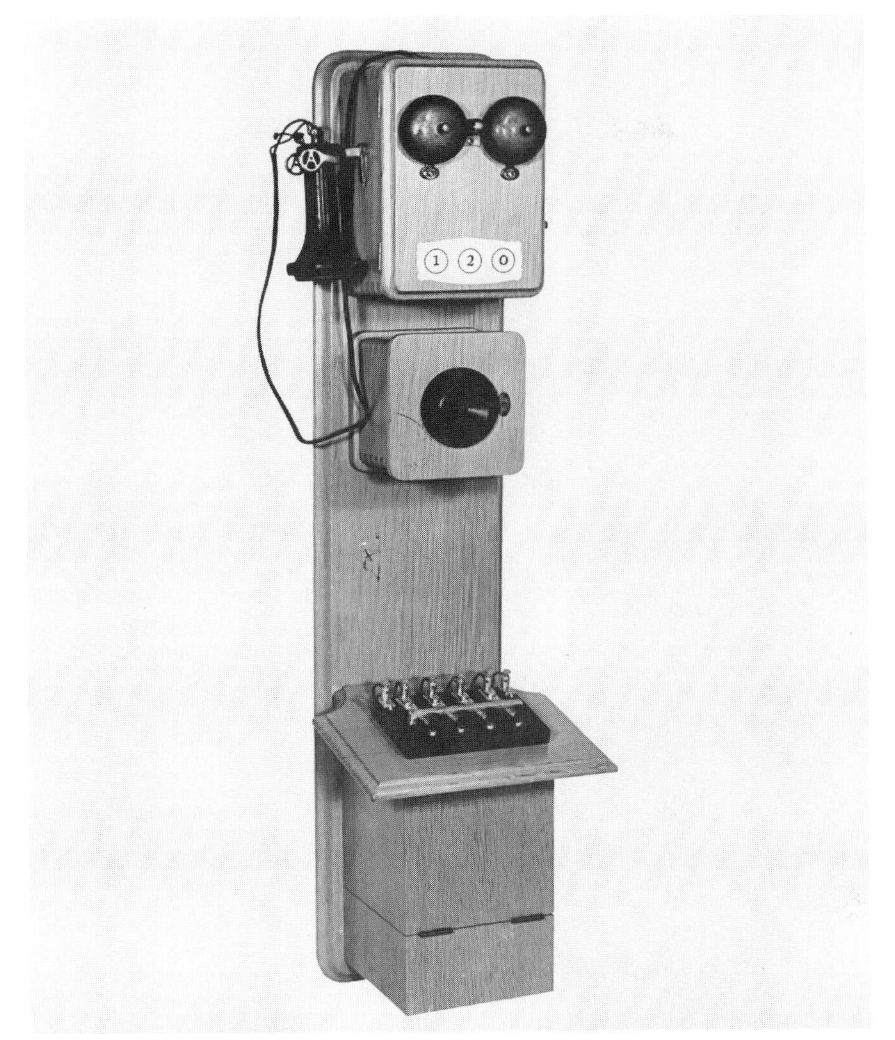

Fig. 6.37 *The first telephone devised by Strowger in 1892 to work with his automatic exchange used press buttons to signal the wanted number*

lifted the receiver. The signal transmitter was then brought under the command of the exchange and made to pass forward the wanted number as and when the exchange was ready to accept it. The levers remained unmoved throughout the call and if the subscriber wished to make a follow-on call to the same number he needed only to turn the crank and pick up the receiver.

With the Lorimer system it was possible to provide an unusual facility. By

Fig. 6.38 *During the early part of the 20th century, Lorimer type automatic exchanges were installed in several countries. Telephones used with this system signalled the wanted number to the exchange with a unique mechanism called a 'signal transmitter' shown in this 1916 photograph*

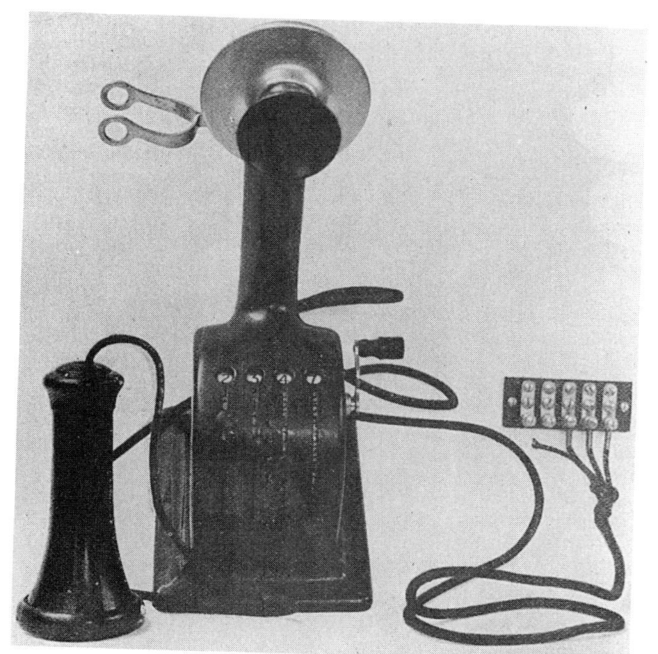

Fig. 6.39 *A Lorimer version of the candlestick telephone with the signal transmitter and crank built into the base. Used at Hereford, Britain's only Lorimer exchange opened on 1 August 1914*

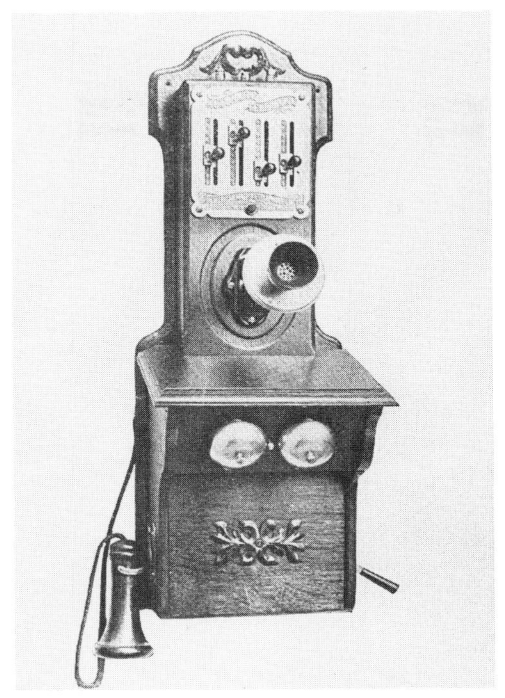

Fig. 6.40 *Telephone made for the Lorimer system by the American Machine Telephone Company c. 1900. The spring of the signal transmitter was tensioned by a lever instead of the more usual crank*

locking the levers of the signal transmitter inside the telephone, calls could be restricted to one predetermined number only. Many years later, equipment was developed to provide a similar facility with other exchange systems. This enabled taxi proprietors, for example, to put telephones in public places so that people could call the taxi office but could not make any other calls.

In the early 1900s, telephones with calling devices superficially similar to the Lorimer signal transmitter were used with a system developed by J.J. Brownrigg and J.K. Norstrom. This system was manufactured by the Globe Automatic Telephone Company and used, on a small scale, in the USA. One interesting claim made by this company was that the exchange equipment could be driven by electricity, gas, compressed air or hydraulics.

In the same period, several systems were developed which used telephones with unconventional dials. One was the Clark Automatic Switchboard system invented by Emery A. Clark, a Bell Telephone Company employee but used, in the main, by other companies. It had a maximum capacity of 75 lines and was intended to provide automatic telephone service in small towns and rural

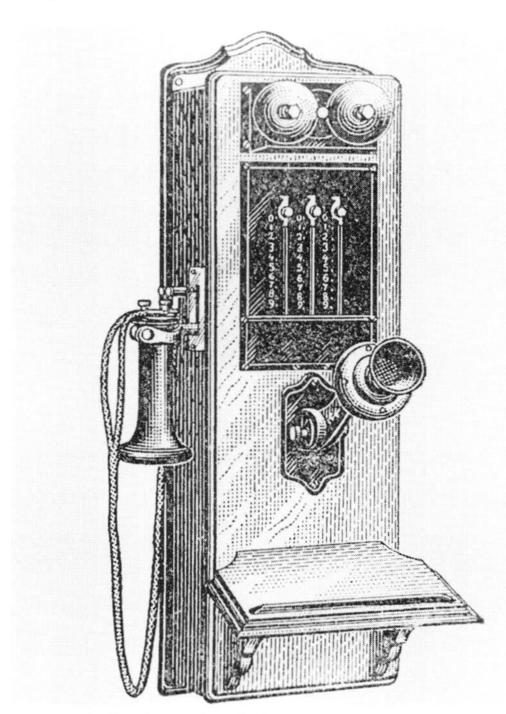

Fig. 6.41 *Instrument made in the early 1900s by the Globe Automatic Telephone Company with calling device controlled by three sliding knobs*

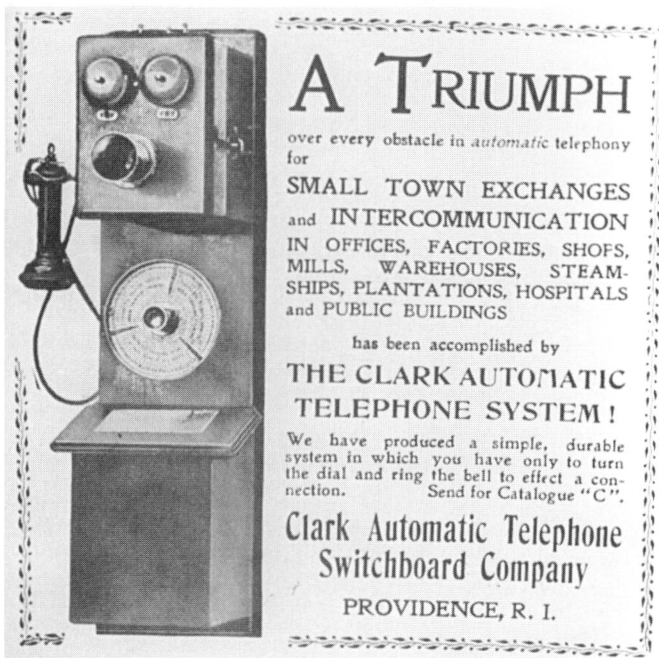

Fig. 6.42 *An advertisement of 1903 for the Clark automatic telephone*

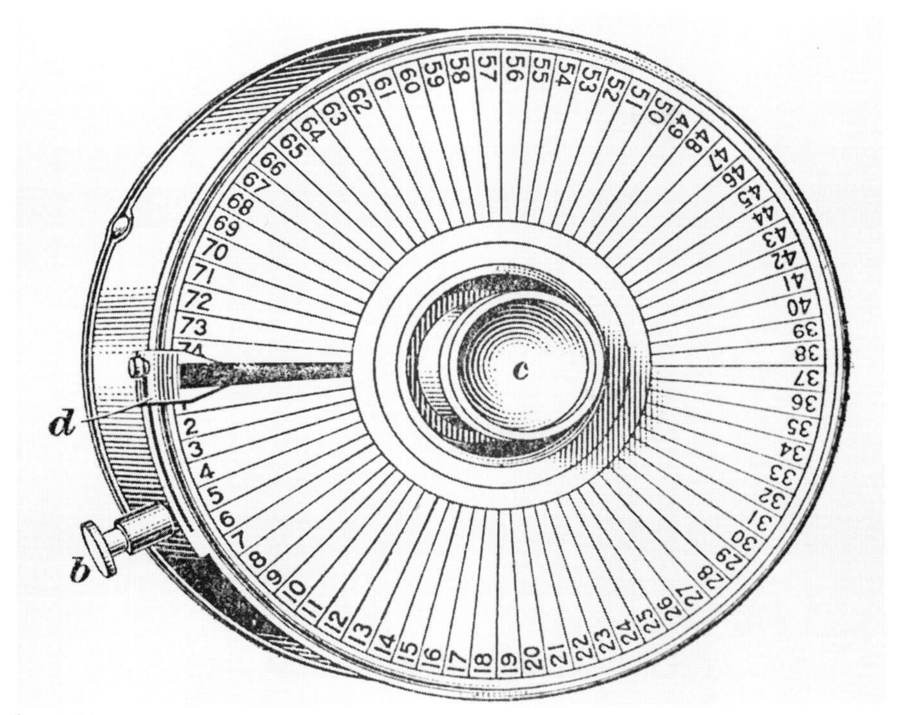

Fig. 6.43 *Detail of the dial used with the Clark automatic telephone, c. 1903*

Fig. 6.44 *Telephone with Western Electric Co. automatic dial. c.1906*

Fig. 6.45 *Telephone with Munson dial used in several places in Minnesota between 1907 and 1912*

Fig. 6.46 *A telephone dial made in 1896 – the year in which the dial was invented. The 'finger flanges', on which the finger was rested, can be seen*

Fig. 6.47 *A Strowger telephone of 1896. The dial had 'finger flanges' instead of finger holes which were introduced the following year*

communities. It was also used for branch exchanges in factories, offices etc. Most of the installations were in north western Nebraska near Sioux City but there were a few in other parts of the USA and even as far afield as Europe.

Various types of telephone were fitted with the Clark dial to enable them to be used with the Clark system. The dial itself consisted of a disk, about the size of a tea plate, with a knob at its centre. The disk was divided into 76 segments, one of which was left blank while the others were numbered 1 to 75. There was also a space on each segment where a name could be written. Although the disk was on a spindle, it was normally prevented from turning by a locking device. This could

Fig. 6.48 *A wall telephone of 1905 made by the Automatic Electric Company of Chicago, USA. The 11-hole Strowger dial had one hole labelled 'long distance' which was used to call the exchange. During the early 1900s this Company exported telephones to most countries where Strowger exchanges were used*

be released by pressing a button at the side of the dial. To make a call, the button was pressed and the knob turned until the wanted number was adjacent to a fixed pointer. The locking button was then released. This relocked the disk and the wanted number was transmitted.

Another unconventional American dial was used with a system developed by the Western Electric Company. It had a maximum capacity of 100 lines and was used for both public and branch exchanges. Around the dial were 100 numbered holes and, in order to give them sufficient space, they were arranged in two concentric circles with the odd numbered holes slightly nearer the centre of the dial than the even numbered holes. An arm which could be rotated over the holes was pivoted at the centre of the dial. There was also a metal peg which could be placed in any hole. To prevent this peg being mislaid, it was attached to the centre of the dial with a chain.

To make a call, the peg was fitted into the hole bearing the number of the wanted subscriber. The arm was then rotated until it was stopped by the peg. It was then released and it returned slowly to its original position under the control

Fig. 6.49 *A telephone of 1905 nicknamed by subscribers the 'pot belly' phone. It was one of the models made by the Automatic Electric Company of Chicago. Telephones made by this firm can be instantly recognised by the Company's hall-mark, the letter 'A' stamped in the switch-hook*

of clockwork mechanism within the dial. To return from '100' took ten seconds and to return from lower numbers took proportionally less time. The object of these holes and the peg was to ensure that the arm was rotated by exactly the correct amount.

Another system which used a dial with holes and a peg was invented by Augustus Munson and installed in several places in Minnesota, USA between 1907 and 1912.

By the end of the first decade of the 20th century the superiority of the Strowger dial was beyond dispute. Other dials, and the systems they were used with, had limited capacity. But the Strowger dial, when connected to a Strowger exchange, was capable of being repeatedly operated – so enabling any number of

Fig. 6.50 *A manual exchange telephone converted, by the addition of a dial, into an automatic exchange instrument, c. 1920*

digits to be transmitted in sequence. This meant that there was virtually no limit to the size of a Strowger system.

The earliest Strowger dials had small projections either from the face or the edge of the dial on which the caller placed his finger when dialling. These projections were called 'finger flanges'. They were quickly superseded by finger holes. However, there was no general agreement on how many finger holes there should be. In Germany, a dial manufactured by Siemens and Halske had no less than 25 finger holes. The most popular number of holes was either 10 or 11. The advantage claimed for the 11 hole dial was that the full numbering range could be used for subscribers while the 11th hole, which was generally labelled 'operator' or 'long distance', could be used to call the exchange. However, experience showed that there was no substantial economic gain and that subscribers were confused when the first digit of a number was '0'. The 11 hole dial was therefore dropped in favour of the 10 hole dial and it became common practice to use the digit '0' to call the operator. Even then there were differences in the way in which the numbers were allocated to the holes. Most dials were numbered in an anti-clockwise direction, starting with '1' and finishing with '0' but there were

Fig. 6.51 *A table telephone made by the Automatic Electric Company of Chicago and used with Britain's first automatic exchange installed at Epsom in 1912*

exceptions. In New Zealand, for example, '0' was in the usual place but holes '1' to '9' were labelled in reverse order.

The earliest dial telephones were adapted manual exchange telephones. Later, some instruments were designed specifically for automatic use, while others were designed so that they could be easily modified for either manual or automatic working.

Chapter 7

First World War and post-war telephones

At the time the telephone was invented, the telegraph was already well established. In addition to its civilian uses, the telegraph had been adopted by a number of armies. In 1858 it had been used in the Crimean War. It was also used in the American Civil War of 1861 to 1865. But, as well as demonstrating its usefulness for military purposes, it had also displayed its weaknesses. It required special skill to operate and, if the telegraphist was killed, the equipment was of little or no use. The telephone, on the other hand, required no skill and very little knowledge to operate. It was therefore quickly recognised as the solution to the problem. As a result, within a few years of its invention, the telephone was in use with the armies of a dozen or more countries.

Telephone manufacturers were quick to provide telephones for their military customers. They produced a number of what were called 'field telephones'. These were essentially portable instruments made in self-contained units and requiring the minimum of installation. They were not made exclusively for military use but were well suited for military needs.

One of the first manufacturers to produce equipment of this category was Siemens and Halske of Germany. By the 1880s this firm was making a field telephone kit comprising a cable drum upon which 550 yards of telephone cable was wound. Stored within the hollow core of this drum were two of the improved Bell telephones designed by Werner von Siemens. As no batteries were needed with these instruments, this was all that was required to install a telephone link in the minimum time. A strap, fastened to the sides of the cable drum, enabled the entire apparatus to be carried by one man.

As new improvements in telephone technology were made, they were applied to field telephones. By the beginning of the 20th century the typical field telephone had evolved into a handset instrument with a granular transmitter. When not in use, the handset was housed in a wooden case which also contained the other telephone components including the battery. Telephones such as this proved their worth in a dozen or more military campaigns. But the greatest test

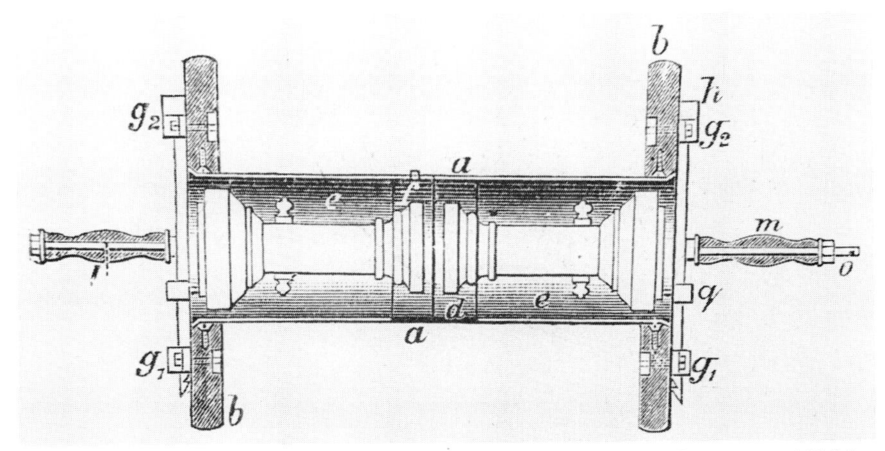

Fig. 7.1 *Drawing showing field telephone kit of the 1880s made by Siemens and Halske of Germany. Two Bell telephones, stored within a hollow cable drum, plus the cable (which is not shown) enabled a telephone link to be set up in the minimum time*

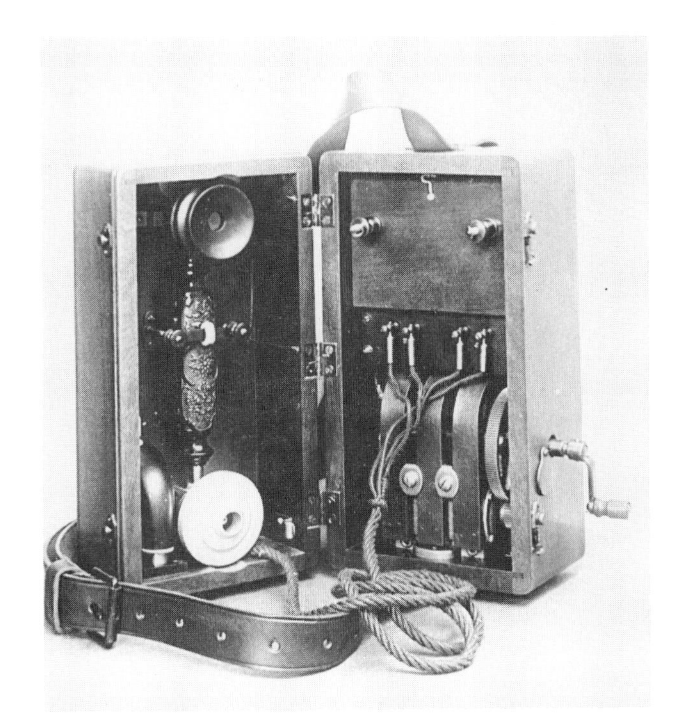

Fig. 7.2 *Field telephone made in Antwerp, Belgium by the Western Electric Company in the early 1900s. The handset, with mouthpiece removed, is fitted into the left hand side of the case while, on the right, there is a calling magneto and a compartment for batteries*

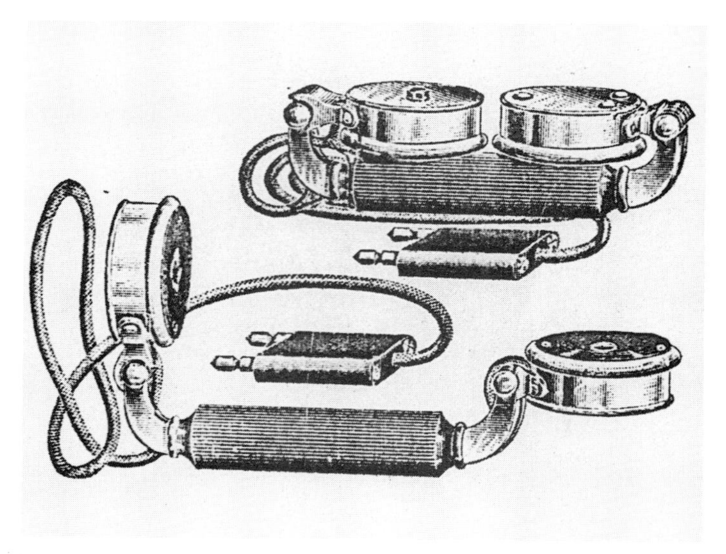

Fig. 7.3 *Folding handset illustrated in the 1903 catalogue of the Western Electric Company of London*

Fig. 7.4 *A British Army telephone of 1908, designated 'Telephone D Mark 1', made by the British L.M. Ericsson Manufacturing Company of Beeston, Nottingham. The hinged flap over the transmitter served two purposes. When the telephone was not in use it was closed to protect the transmitter, and when the telephone was in use it was opened to form a mouthpiece*

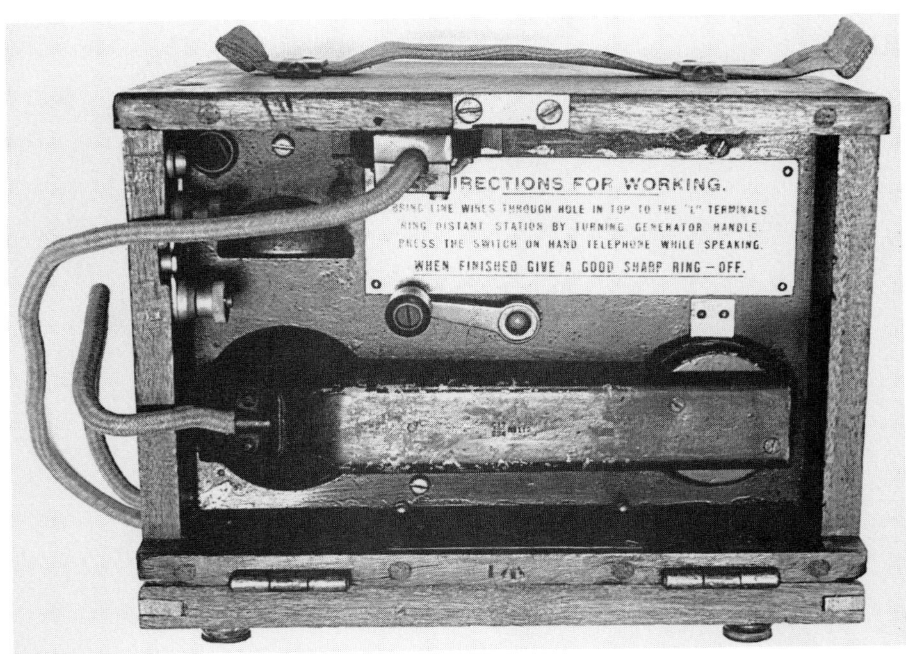

Fig. 7.5 *One of the 42,000 field telephones manufactured by the British Post Office during the First World War and used by the British Army on all fighting fronts. It was commonly called a 'Trench telephone'*

Fig. 7.6 *A complete telephone with all the components, including a calling buzzer, contained within the handset. Originally produced by Ammon in 1905 and used by the German Army in forward positions throughout the First World War. The mouthpiece folds flat with the body when not in use*

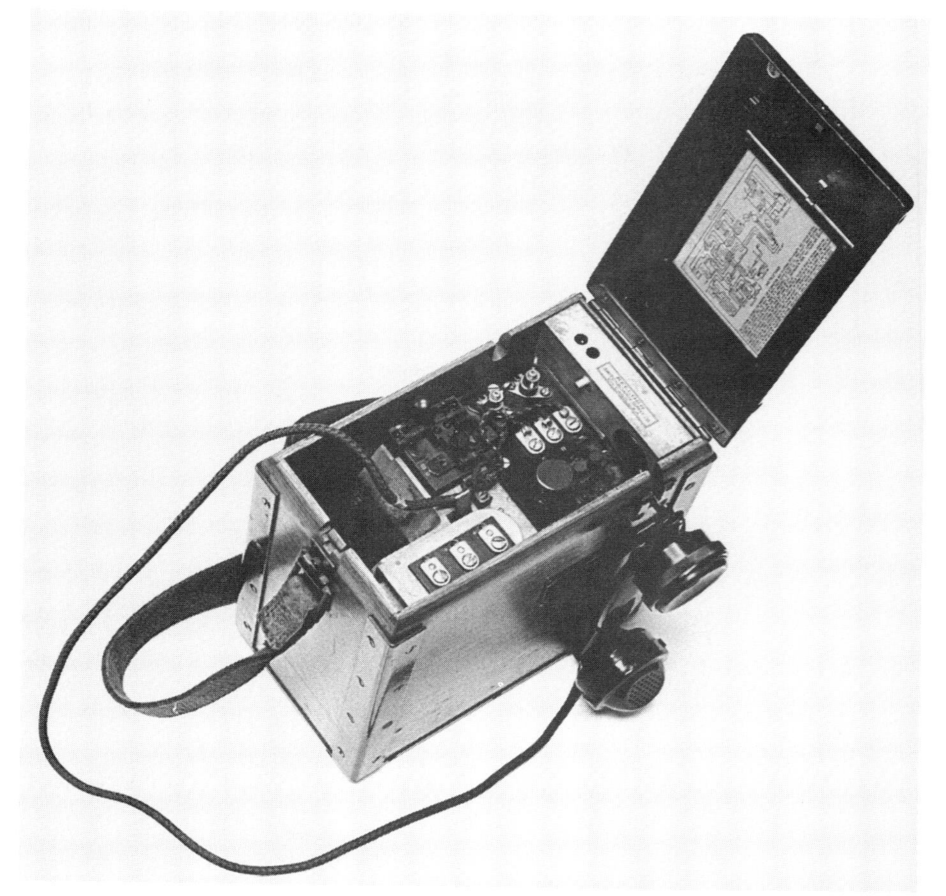

Fig. 7.7 *United States Army field telephone, used during the First World War, manufactured by the Kellogg Switchboard and Supply Company of Chicago. When in transit the handset is stored inside the case*

came with the First World War. In the static trench warfare which developed in Flanders there was a great need for good communication. Telephone lines, often buried at considerable depth, linked various parts of the front line with each other and with the base areas. The telephone exchange at Allied GHQ had direct lines to London and Paris and would have been large enough for a medium sized town.

In forward areas, telephone lines were repeatedly damaged by shell fire and other causes and there was considerable leakage of current into the earth. Even greater earth currents were caused by telegraphs because, in many cases,

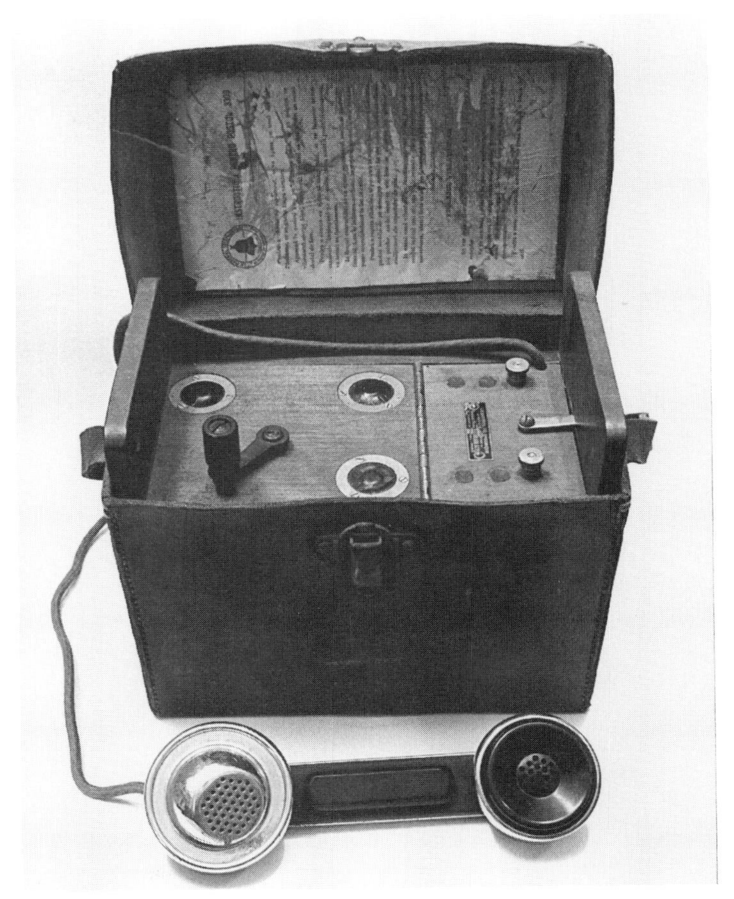

Fig. 7.8 *A French field telephone made by Aboilard of Paris utilising a handset made by the Western Electric Company*

telegraph lines consisted of a single wire, which was used to carry the outgoing current, while the returning current was deliberately passed through the earth. The German signallers discovered that by burying metal plates as near the Allied lines as possible, these earth currents could be intercepted and Allied messages read. To make the intercepted currents stronger, the German signallers used electronic amplifiers and thus was born the practice of what later came to be known as 'electronic surveillance'.

The Allies discovered their messages were being intercepted and took countermeasures. At first they attempted to obliterate the earth currents their signals produced, by using powerful buzzers connected to plates buried in the

Fig. 7.9 *First World War postcard showing a soldier ringing home but, at that time, for most soldiers, ringing home was no more than a dream*

earth. The high-intensity currents these buzzers created must have made the job of the German signallers more difficult. But they also found their way into Allied telephone circuits and could be heard, with varying degrees of loudness, on telephones all along the front.

A second, and more sophisticated approach to the problem was the adoption of a secret telegraph system called the 'Fullerphone'. It was named after its inventor, Captain (later Major General) A.C. Fuller of the British Army's Royal Engineers. This telegraph used an ingenious circuit which balanced the current in such a way that any leakage from the line produced a continuous buzz which was meaningless to an eavesdropper.

Fig. 7.10 *Fullerphone-telephone set made by the British Post Office during the First World War*

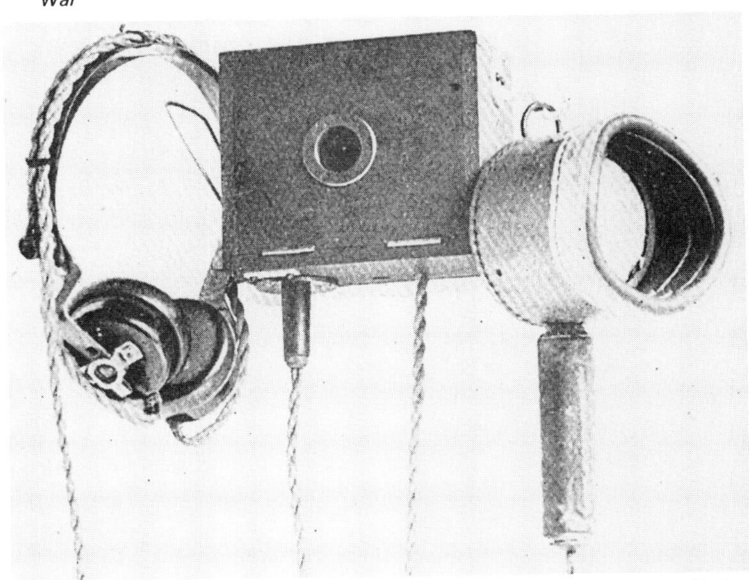

Fig. 7.11 *Airship telephone set with double earphones and transmitter mounted in leather face mask to exclude background noise. Telephones of this type were used on board the British R34 which became the first airship to cross the Atlantic when, in July 1919, it crossed in both directions*

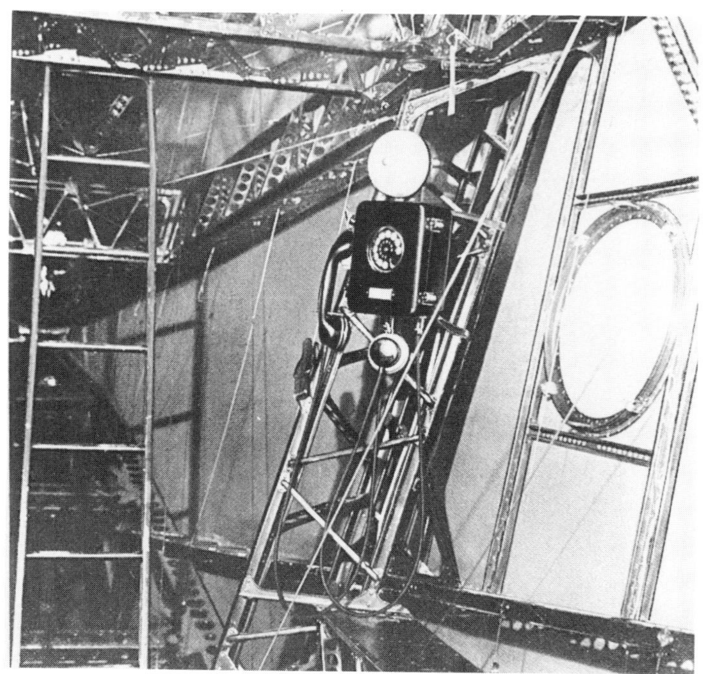

Fig. 7.12 *With engines housed in external pods, the lower noise level within the German airship Graf Zeppelin LZ 130 enabled conventional handset telephones to be used. In this 1938 photograph a Siemens & Halske telephone is pictured amongst the lattice spars and control wires in the airship's envelope. Telephones were interconnected with a small automatic exchange*

Several thousand Fullerphones of various designs were constructed for the Army by the British Post Office from parts originally intended for telephones. Later, Fullerphones were made by several telephone manufacturers.

Fuller showed that his secret telegraph could be conveniently superimposed on an ordinary telephone circuit. When used in this way the secrecy of the telegraph was unimpaired, although telephone conversations were not secret. The Fullerphone-telephone combination proved so useful that it became normal for the two circuits to be incorporated in a single unit and Fullerphones without telephones became extremely rare.

Quite a different war-time problem was communication in airships. In a typical airship, members of the crew were suspended in a number of widely separated gondolas. Movement from one part of the ship to another was difficult even if the crew could leave their posts. This situation produced a very real need for communication and the telephone was the obvious answer. However, as in some industrial situations, special problems existed. Airships used hydrogen

Fig. 7.13 *Aviation headphone and throat microphone set developed during the First World War to combat background noise in aircraft. The throat microphone was unresponsive to air-borne noise and was operated directly from the vibrating larynx*

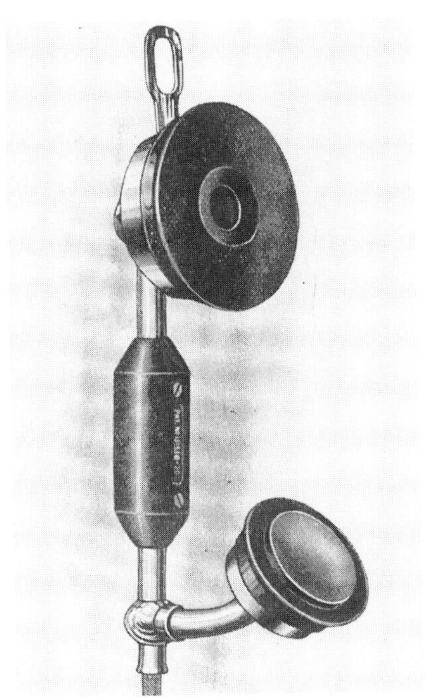

Fig. 7.14 *Laryngaphone handset for noisy situations comprising a conventional telephone receiver and a throat microphone. This combination was originally developed during the First World War for use in airships*

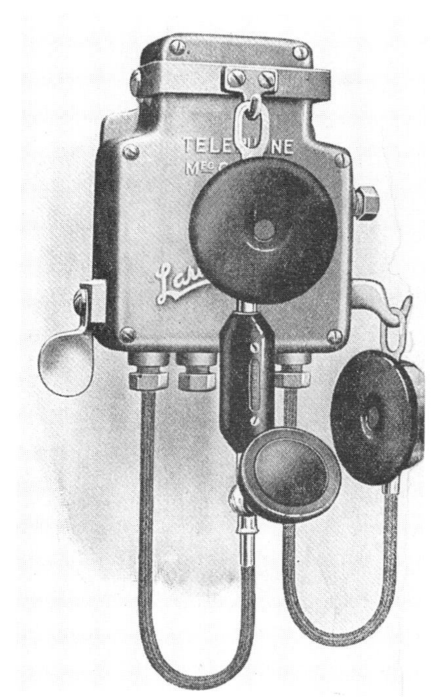

Fig. 7.15 *Laryngaphone wall telephone set with additional earpiece. Based on technology developed during the first World War and adapted for noisy industrial situations by the Telephone Manufacturing Company of Dulwich, London. The metal telephone case was dust-proof, watertight and non-corrodible.*

and there was always the possibility of an explosive atmosphere. Telephone equipment therefore had to be airtight. A further problem was engine vibration which telephones had to be capable of withstanding. But, by far the greatest problem was noise. The engine gondolas were the noisiest places, and it was not unusual for the engineer to be penned in close proximity to engines producing 1,000 horsepower or more.

The first solution to the noise problem was to use a telephone set with double earphones and a transmitter in a special mounting consisting of an oval leather-covered box, open on one side, and cushioned at the edge so as to fit the face. This arrangement helped to exclude some, if not all, of the background noise.

Later, airship crews benefited from the knowledge gained from a more generalised investigation into the problem of using telephones in noisy situations. This had commenced even before the First World War as a result of a discovery made by telephone users. Subscribers with candlestick telephones who worked in noisy places, discovered that they could be heard better if, instead of holding their transmitters in front of their mouths, they held them to their chests.

Fig. 7.16 *A British candlestick telephone of the 1920s. This instrument was so much like its American counterpart that a casual observer would be unlikely to notice the minor differences*

There were experiments on both sides of the Atlantic to assess the practicability of a chest transmitter. However, it was soon discovered that better results could be obtained by placing the transmitter against the neck on either side of the windpipe. From this discovery, throat transmitters were developed. In airships and observation balloons, a special handset was used which combined a conventional telephone receiver with a throat transmitter.

In aeroplanes, which in those days had open cockpits, aviators wore helmets which incorporated telephone receivers, while around their necks they wore a throat transmitter attached to what was called a 'necklet'. Both the handset and the helmet-necklet combination proved successful as a means of transmitting the speaker's voice while excluding unwanted airborne noise.

After the War, the British Post Office adopted a series of telephones with throat transmitter handsets. This series included wall and table models for automatic and manual exchanges of several types. They were intended for use in noisy situations and were offered to the public under the name 'Laryngaphone'. This name was derived from the word 'larynx', the part of the throat against

Fig. 7.17 *This wall telephone, manufactured in Britain in the 1920s, was a less ornate version of a National Telephone Company instrument of 1900*

which the transmitter rested. They were never popular and were quickly withdrawn.

Laryngaphone transmitters attached to necklets similar to those worn by airmen, were successfully used by divers and did much to reduce the problem of background noise caused by air entering and leaving the helmet.

In Europe, the First World War was a watershed in innumerable ways. When peace returned, practically everything was changed. Gone were the customers who could afford telephones made with precious metals, as some of the 'golf ball' candlestick instruments had been. Gone too, were customers who could afford telephones made by teams of the finest craftsmen, as was the case with the early L.M. Ericsson instruments. In future, telephone administrations, in common with many other businesses, would have to seek their new customers from the price-conscious masses. Businesses in general were faced with the problem of producing attractive goods at a price the man in the street could afford. It was apparent that the way to keep prices down was to use machines, but, if machine-

Fig. 7.18 *The resemblance between this American telephone of the early 1900s and its British counterpart was very close and one of the few differences was the type of wood used. Judging from the telephone user's smile, the sweet words the telephone carried were probably similar too!*

Fig. 7.19 *French telephone of 1917 made in the style inspired by classical architecture which was popular with European telephone manufacturers before the First World War*

Fig. 7.20 *The pedestal telephone, popular in France in the 1920s, was a cross between European handset telephones and the American candlestick. This instrument was made by Thomson-Houston of Paris in 1920*

made goods were to be attractive it was important that designers should take into account the capabilities and limitations of machine manufacture. Artists and designers accepted this challenge. A new artistic movement came into being in which all types of goods were designed specifically for machine manufacture while decorations were often inspired by the working parts of the machines themselves. This movement had different names in several countries but is probably best known as 'art deco'. This is a drastic abbreviation of the name of the exhibition where designs inspired by this artistic movement were first displayed – 'L'éxposition Internationale des Arts Décoratifs et Industriels Modernes' – which took place in Paris in 1925. The impact of art deco on telephone design was considerable and it would be difficult to imagine any representative exhibition of art deco objects which did not include telephones.

In North America, where there were fewer traditional class distinctions, the social upheaval following the First World War was less drastic. From its invention, the telephone had served a broad spectrum of American society and,

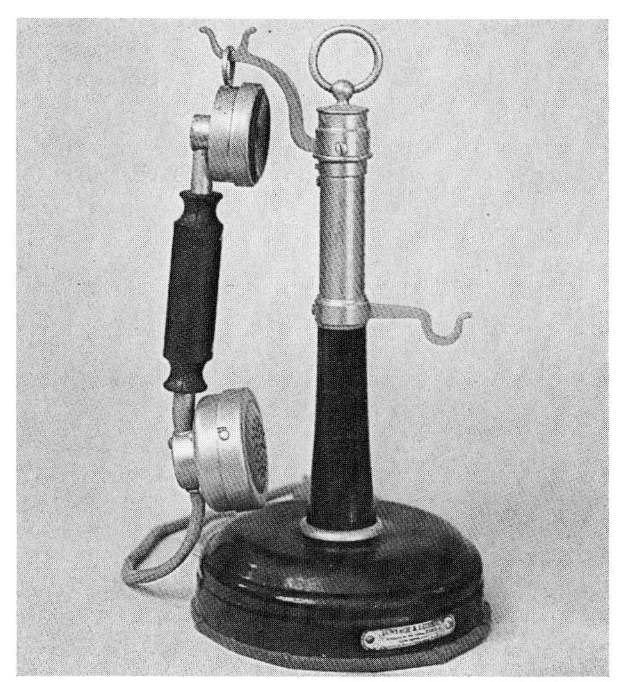

Fig. 7.21 *A French pedestal telephone of 1920 made by Duniach and Leclerc. A carrying loop is attached to the top of the stem*

Fig. 7.22 *A pedestal telephone made by Grammont of Paris in 1920. The simple elegant lines of the wooden stem reveal the influence of the art deco movement*

Fig. 7.23 *This British telephone of the 1920s is apparently for manual exchange use only . . .*

Fig. 7.24 *. . . but when the notice frame is removed the fixing for a dial is revealed*

Fig. 7.25 *Fit a dial and make a few internal connections . . .*

Fig. 7.26 *. . . and the metamorphosis is complete. What had been a manual exchange telephone can be used for dialling calls*

Fig. 7.27 *This telephone was made by a number of manufacturers to a standard British Post Office specification and used in a number of countries from the early 1920s*

Fig. 7.28 *Telephone made by Siemens Brothers of London for South Africa's first automatic exchange installed at Waterkloof, Pretoria, and opened on 19 July 1922. The South African emblem appears above the dial*

Fig. 7.29 *Candlestick telephone and bell manufactured by the Western Electric Company, USA, in 1920. This was the first type of table telephone used when the automatic system was introduced in the city of São Paulo, Brazil, in 1928*

in the main, American telephones had always been down-to-earth utilitarian instruments which were completely appropriate in the post war world.

In Britain, telephones had always been strongly influenced by their American counterparts and, in the period following the War, British and American telephones became so much alike that in a number of cases, a casual observer would have been unlikely to notice the differences.

Compared with Britain, telephone design in mainland Europe was less influenced by American practice. However, one type of telephone, popular in France, was the pedestal handset instrument. This was an interesting mixture of the American candlestick and the French handset telephone. It was made in a variety of designs by several companies in the 1920s and 1930s.

A further change took place about the time of the First World War. Before 1914, a topic for discussion amongst telephone men throughout the world was the relative merits of manual versus automatic exchange systems. In the USA, the independent, that is non-Bell, companies had shown considerable enthusiasm for automatic systems. The Bell companies, by contrast, had considered that this

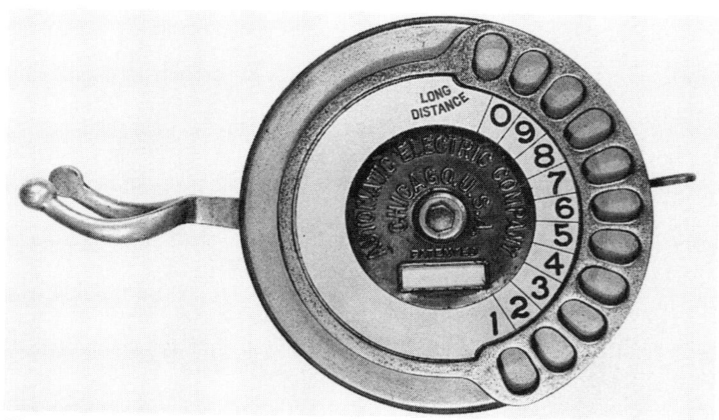

Fig. 7.30

Fig. 7.31

Fig. 7.32

Fig. 7.33

Variations of the Strowger dial

Fig. 7.34

Fig. 7.35

Fig. 7.36

Fig. 7.37

Fig. 7.38

Fig. 7.39

Variations of the Strowger dial

enthusiasm was unjustified and it was not until 1919 that AT&T, the Bell parent company, announced its intention to commence a long-term programme to convert its exchanges to automatic working.

In Britain, the National Telephone Company adopted the same policy as AT&T and had no intention of installing automatic exchanges. But, in 1912, the Company was taken over by the Post Office and, in the same year, the first of a number of different types of automatic exchange was installed to determine which would be the most suitable for use in the United Kingdom.

A similar story, with variations, could be told about a number of other countries. The outcome was that, whereas in 1914 the superiority of automatic exchanges was open to doubt, in the years following the War it was apparent, at least to telephone men, that the automatic exchange was the system of the future. So that even though the overwhelming majority of telephones were still connected to manual exchanges, it became increasingly common in the 1920s for new telephones to be designed so as to be readily converted to automatic exchange working.

Chapter 8

Into the modern era

Although the majority of telephones are connected to public exchanges, many have always been connected to private systems. Such systems vary in complexity from a telephone at either end of a line, to town-sized automatic exchanges providing all the facilities associated with a public service, plus special services such as the recording of dictation.

Telephone instruments used with private systems may be the same as those connected to public exchanges but many special instruments have also been developed. Private telephones are often used within a building, or some other restricted area, and lines are generally short. For this reason, complex circuits and high quality components are sometimes unnecessary and simple, more cheaply-made, equipment can often give good results. The shortness of the lines also makes it economical to interconnect telephones with multi-wire cable instead of the basic pair of wires used with public systems. This enables the bell in each instrument to be connected to an individual bell wire in the cable which, in turn, makes it possible to have a simple but effective form of selective ringing.

New types of coinbox telephone were made necessary by the introduction of automatic exchanges. The earliest automatic exchange coinboxes closely resembled the earlier manual exchange types and, very often, the only superficial difference was the addition of a dial. In the USA, coinbox telephones were being used with automatic exchanges in the early 1900s.

In Britain, automatic exchange coinboxes were brought into use by the Post Office in the 1920s but did not come into widespread use until the 1930s. These boxes were manufactured by Hall Telephone Accessories Ltd., and are of special interest because they were adopted by a number of telephone administrations as widely separated as Australia and Malta. The typical Hall box was fitted with two buttons labelled 'A' and 'B'. To cater for various currencies, up to four slots could be fitted but generally fewer were used. In Britain, for example, there were never more than three.

The way in which the box was used, taking Britain in the 1930s as an example, was first to insert two pennies. The weight of the pennies operated a switch which enabled the dial to be used, but prevented speaking. If the correct number

Fig. 8.1 *A simple, but effective, private telephone made by the General Electric Company of England, c.1890*

answered, the caller pressed button 'A'. This caused the money to drop into the cash container and the switch was released, enabling the caller to speak. If the wanted number was engaged, or if the wrong number was connected, the caller pressed button 'B'. This caused the money to be refunded. It also disconnected the line so that the call was released.

As well as pennies, British boxes had slots for sixpences and shillings (12 pence) but these slots could only be used when calls were made via the operator. The British Post Office started to phase out the Hall coinbox in the 1960s but some were still in use in the 1980s. Those in use in 1971, when decimal coinage was introduced, were converted to accept the new two and ten penny coins.

There was an interesting variation of the Hall 'A' and 'B' button box which had one button only. It was used in a few places in Australia. The function normally performed by the 'A' button took place automatically in response to a signal from the exchange when the call was answered. The single button caused money to be refunded if there was no reply. There was no provision for a refund if the caller was connected to a wrong number. Boxes with no buttons were tried but were not used to any great extent.

In Sweden, a coinbox was developed by L.M. Ericsson in the 1930s with a coin

Fig. 8.2 *Telephone for a ten-line internal system, with selective ringing, made by Elektrisk Bureau of Norway in the early 1900s*

guide, or track, mounted on top of the box. If there was no reply, the caller was able to remove the coin from the guide.

The 1930s was also a period when new types of police telephone were introduced. In Britain, particularly in London, there were police kiosks, the smallest of which was about the size of a telephone kiosk, and the largest, several times that size. One type, in particular, is now famous as a result of the television series 'Dr Who'. These kiosks were, in effect, minor police stations with a desk, and a telephone connected to the main police station. In 1934, new police telephone equipment was designed and a small door was fitted in the kiosks to enable the public, as well as the police, to use the telephone. The idea was successful and, in 1937, the Post Office developed a police pillar which could be used in places where there was insufficient space for a police box. This pillar had many of the features of the earlier Digeon apparatus used in Paris.

In some cases, police pillars were surmounted by an electric light generally called a 'beacon'. It was actuated from the main police station and used to attract the attention of the local policeman when he was wanted. The upper part of the pillar was three-sided, with a door on each side. One door had no lock and could be used by the public. Behind this door was a loudspeaking telephone. The

Fig. 8.3 *A two-dial telephone made by Siemens and Halske of Germany in the late 1920s. The 25-hole dial was used for dialling into a private system and the 10-hole dial was used for dialling into a public exchange*

second door was used by the police; it concealed a telephone handset and a small folding writing shelf. The third door was used for maintenance purposes. Further down the pillar was a fourth compartment which contained a first aid kit.

The pace of technical innovation is seldom steady. For many years there may be only minor advances, then two or three major discoveries may follow in quick succession. In the case of the telephone, periods when development has been slow have been few and far between and, for the most part it has varied from steady to extremely rapid. One period of especially rapid development was in the 1920s. The decade opened with a major advance when, in 1920, G.A. Campbell of the USA announced his invention of a new telephone circuit. Before that date, the

Fig. 8.4 *Advertisement dated 1929*

current from the telephone transmitter passed through the associated receiver causing the telephone user to hear his own voice. This was called 'side tone' and was undesirable because it caused the telephone user to lower his voice in an unnatural way. Campbell's circuit greatly reduced side tone and increased efficiency. It used a new component called an 'anti side tone induction coil'. This name was shortened to its initial letters 'ASTIC' – a term that was also used to describe telephones incorporating this device. So, instruments in which side tone had been virtually eliminated were said to be 'ASTIC' while other telephones were said to be 'NON ASTIC'.

Fig. 8.5 *Gray Pay Station Company coinbox used in conjunction with a 'Pot belly' telephone with an eleven-hole dial. Used in the USA in the early 1900s*

Another development which had a profound and lasting impact on telephone design was the evolution of plastics. For many years telephone manufacturers had felt the need for materials from which telephone components could be easily and cheaply made. Ebonite, a form of hard rubber, had proved useful and had served many purposes. Bakelite had been discovered many years earlier but it did not come into widespread use until the 1920s when improvements in chemical engineering and moulding techniques made it practical. In the second half of the 1920s it became the first material from which a completely moulded telephone was made. Finally, at the end of the decade, urea formaldehyde plastic became available. This material is almost colourless and, by adding pigment, it can be made any colour.

A further advance in the 1920s, was a major improvement in transmitter design. This resulted from work on both sides of the Atlantic. In the USA, the ground was prepared at the beginning of the decade when the Bell Company reconsidered its attitude to handsets. The problem was that due to movement, transmitters in handsets were marginally less efficient than fixed transmitters. In spite of this, handsets had been successfully used in Europe for a number of years. However, in Europe, slight loss of efficiency was less important than in North

Fig. 8.6 *Gray Pay Station Company coinbox used with a wall-type telephone with an eleven-hole dial made by the Automatic Electric Company of Chicago. Used in North America around 1908*

America where calls over distances of 3,000 miles and more were made. In the days before amplification, calls over such distances were achieved only with the aid of the best available transmitters, finger-thick line wires and the lung power of the telephone user. After electronic amplification came into widespread use in the 1920s, the situation changed. It was no longer essential to get the utmost power from a transmitter and the loss of efficiency caused by mounting transmitters in handsets would be acceptable if a transmitter could be devised which gave the same response in any position. As a result, a research programme was conducted by the engineers of the Bell Company's principal supplier, the Western Electric Company. This culminated in 1927 with the production of a telephone handset acceptable to the Bell Company.

The new handset was used with a telephone that was clearly derived from the candlestick instrument. Its lower part resembled a candlestick telephone which had been chopped off a short distance up the stem. Attached to the top of this shortened stem was a cradle which supported the handset in a horizontal position. When it was made available to the public the favourable response was beyond expectation. Northwest Bell, for example, anticipated orders for 3,000

Fig. 8.7 *An American street telephone of the early 1920s. A coinbox instrument was inside the box and a directory hung on the inner side of the door*

instruments during the first year. In the event, 10,000 orders were received. The public quickly nicknamed the new instrument the 'French' telephone. This was because many Americans had seen handset telephones for the first time while serving with the army in France during the First World War. Handset telephones of later design were also sometimes called 'French' but no telephone has a better claim to the nickname than the first Bell model.

In Britain, the telephone manufacturer Siemens Brothers of Woolwich, London, in conjunction with the British Post Office, commenced research about 1924 to overcome the defects in telephone handset transmitters. This research also culminated in the production of a transmitter which, although fundamentally different from its American counterpart, was equally successful.[30]

Fig. 8.8 *A public telephone designed to be fitted to a telephone pole. Used in the U.S.A. in the 1920s. The manufacturer advertised it as being 'especially useful to automobilists who have occasion to communicate with home or garage'*

Siemens Brothers designed an all-moulded telephone to incorporate the new transmitter. It was called the 'Neophone'. Its shape was said to have been inspired by an Edwardian inkstand but it was far from old-fashioned. Indeed, it was very advanced for its time and its well-balanced proportions made it one of the most attractive telephones ever produced. It was originally moulded in black bakelite but, shortly after its introduction, urea formaldehyde plastic became available and it was moulded in that material. Using pigmentation, the telephone was produced in red, green, ivory and mottled brown. There were also versions in gold and silver although these were made by surface treatment of the components before assembly. The Neophone had no bell within its case. Originally the bell was accommodated in a wall-mounted wooden box. Later, bell cases were moulded from the same materials as the telephones themselves and could either be wall-mounted or fitted as a plinth to the bottom of the telephone.

Fig. 8.9 *A typical British public telephone installation used in manual exchange areas in the early 1930s. The kiosk was designed by Sir Giles Gilbert Scott, RA who also designed the Anglican Cathedral in Liverpool*

The idea of an 'add-on' bell case was later used with other telephones but it is difficult to design an instrument that has pleasing proportions both with and without the bell case. The tendency is for the bell case to make the telephone appear 'bottom-heavy'.

The Neophone was adopted by the British Post Office as one of its standard models and it was used by several overseas administrations. Variations of the telephone were also made by other companies. It continued to be one of the standard instruments of the British Post Office until the 1960s and its long active life is indicative of the satisfactory service it gave.

It was the first telephone to be made entirely in plastic and at the time it was conceived, in the late 1920s, the art of designing for manufacture in plastic was in its infancy. It is easy to be wise after the event, and see that the potential of the new material was not fully exploited. Nevertheless, it must be said that the methods of construction used were more appropriate to metals than to plastics. For example, the telephone case was assembled from a number of mouldings and joined with metal screws. The failure to exploit the potential of plastic probably increased manufacturing costs but, fortunately, did nothing to detract from the telephone's attractive appearance.

On the morning of 8 September 1934, a ship, the 'Morro Castle', was destroyed

Fig. 8.10 *Made about 1930, this version of the coinbox manufactured by the Hall Telephone Accessories Company incorporated a rest for a Bell receiver. Later, handset telephones were used with this coinbox and boxes without the Bell receiver rests were manufactured*

Fig. 8.11 *A later version of the Hall coinbox used with a handset telephone. The black and chromium fittings were introduced by the British Post Office in 1936*

by fire off the New Jersey coast of the USA with the loss of 134 lives. The subsequent investigation revealed that the ship's telephone system had completely failed and this was judged to be a major contributory cause of the high loss of life. In particular, it was concluded that many lives could have been saved if there had been a working telephone between the bridge and the engine room. As a result of this investigation, the US Bureau of Marine Inspection and Navigation was convinced that all types of telephone then in use on board ships were inadequate. Central battery telephones were vulnerable to loss of power from the central source. On the other hand, telephones with individual batteries were frequently found to be out of order because the batteries had not been maintained. There was an obvious need for a system that would be ready for use at any time and was not dependent on batteries.

Fig. 8.12 *A Copenhagen street telephone with facilities for making emergency fire, police and ambulance calls. Introduced by the telephone administration K.T.A.S. in 1931*

Strangely enough, such a system had existed, in an elementary form, since the telephone was invented. Bell's original electromagnetic telephone had used no batteries. As a result of the increased technical knowledge that had been gained since Bell's early experiments, it was possible to produce a greatly improved version of his telephone. In its modern form this was called a 'sound-powered' telephone.

Even before the 'Morro Castle' disaster, sound-powered telephones were used on board ships but, as a result of the Bureau's investigation, regulations were issued making them mandatory. Similar regulations were issued by other countries governing their own shipping.

A testing time came with the Second World War when sound-powered telephones, fitted in both merchant and naval vessels, proved their worth under the most arduous conditions.

The major breakthrough in moulded telephone manufacture came in 1931 when Elektrisk Bureau of Oslo, Norway produced a design which exploited to

Fig. 8.13 *British Police Kiosk showing a member of the public using the Loud-speaking telephone to contact the main police station. 1939*

the full the mouldability of plastic materials. Unfortunately it failed to achieve the Neophone's attractive appearance. Fundamentally it was an inverted box with its upper part shaped to form a resting place for the handset. This simple idea of a one-piece case has since been used dozens or even hundreds of times in every country where telephones are made. For sheer practicability it has never been bettered and today, it is embodied in so many telephones that it seems strange that there was a time when the idea needed to be thought of. It was, perhaps, unfortunate that the early 1930s was a period when the design of everyday objects was strongly influenced by the Cubist movement. This probably

Fig. 8.14 *A police pillar in an English street in the 1930s showing the door used by the public*

explains the telephone's hard lines and sharp edges which few people admire today, although, no doubt, they looked modern and fashionable at that time.

The Elektrisk Bureau telephone was developed by L.M. Ericsson of Sweden, an associate of the Norwegian company, and almost immediately, the Swedish telephone administration, Televerket, adopted it as its standard model. At this point in the history of the instrument, design and technology took second place to fate in determining its future. In October 1932, the Prince of Wales, the eldest son of King George V of Great Britain, visited the Stockholm exhibition where the telephone was displayed. He admired it and selected it for use in his home. The approval of this fashionable man-about-town did much to promote the telephone's success beyond Scandinavia.

The British Post Office was already using the Neophone, but wished to include in its range a telephone with integral bell. This was one of the features of the Scandinavian telephone and the Post Office, working in conjunction with Ericsson Telephones Limited – the British associate of L.M. Ericsson – developed a composite telephone. This used the case of the Scandinavian instrument in conjunction with the Neophone handset and new internal equipment. Some telephones were also fitted with the drawer first fitted in the Neophone. This was

Fig. 8.15 *A policeman using the telephone and folding writing shelf in a British Post Office police pillar in the 1930s*

intended for a list of dialling codes and personal telephone numbers. Several versions of this telephone were made for different types of exchange and they were manufactured in black, red, ivory and green. These telephones were allocated type numbers in the 300 range and were collectively referred to as the '300 series'. They came into general use in 1937. Several manufacturers made considerable quantities of these telephones to British Post Office specifications. They were also exported to Australia, India, New Zealand, South Africa and many other countries.

In 1937, the Bell Telephone Company of America adopted a telephone made in metal. However, its shape was suitable for production as a one-piece moulded plastic case and, from 1940 onwards, it was made in this way. The manufacturer was the Western Electric Company. Although there is no evidence that it was inspired by the Scandinavian instrument, it also had hard lines and sharp edges. It was manufactured only in black and today it is doubtful if anyone, except a dedicated telephone enthusiast, would want one as an ornament. Like its British counterpart, it was also given the type number '300' and was generally referred to as the '300 type desk set'.

The independent telephone companies in North America, that is the

Fig. 8.16 *Telephone manufactured by Thomson-Houston and used extensively in France. Produced in dial version, 1924*

Fig. 8.17 *The first handset telephone adopted for general use in North America. Designed by the engineers of A.T.&T. and the Western Electric Company and introduced in 1927. First World War veterans nicknamed it the 'French' telephone*

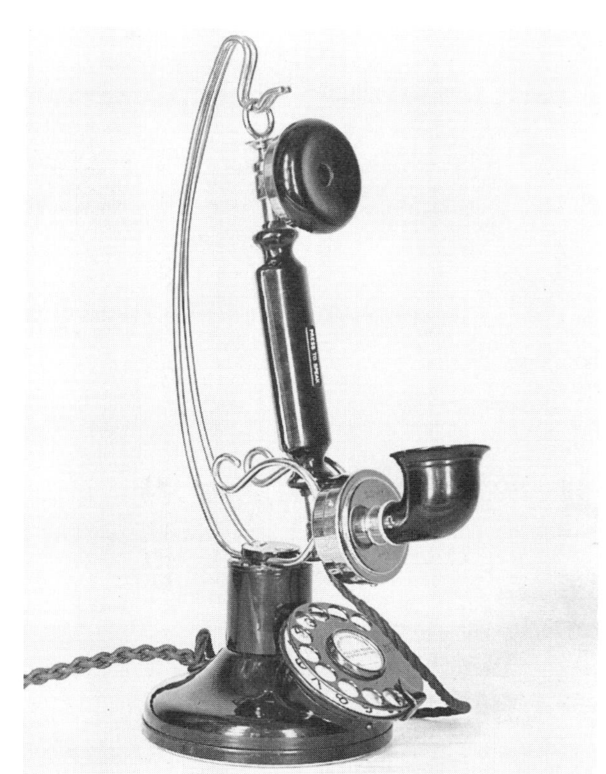

Fig. 8.18 *This telephone, made by Siemens Brothers of Woolwich, London, in the mid 1920s was clearly derived from the candlestick instrument. It was used by the British Post Office but the rapid pace of technological development condemned it to a short life*

Fig. 8.19 *The Neophone, introduced by Siemens Brothers of Woolwich, London, in the late 1920s was one of the first moulded telephones*

Fig. 8.20 *Some 1930s versions of the Neophone had a drawer in the base which contained a hinged cellulose acetate holder for a dialling code list. It was not unusual for housewives to put a currency note in this drawer so that it would be at hand to pay a household bill. But there was no partition between the drawer and the inside of the telephone so, when the drawer was opened the note was frequently found to have vanished. It did not re-appear until the next maintenance visit*

companies not belonging to the Bell consortium, also adopted telephones with one-piece moulded cases. These instruments were made by Automatic Electric, Kellogg, North Electric, Stromberg Carlson and other manufacturers. Wall telephones with one-piece moulded cases were also made by these companies. The Automatic Electric Company used the name 'Monophone' to describe its range of handset telephones. However, this term had long been used in Europe to describe a particular type of handset which used a horn to convey speech to a transmitter mounted behind the receiver.

European manufacturers also made telephones with one-piece moulded cases. L.M. Ericsson produced a wall telephone of this type in 1931 – the same year in which their one-piece table telephone originated. Both telephones were in the same style and, if anything, the wall telephone had even harder lines than those of the table instrument. At that time, the British Post Office had a wooden wall telephone which met its requirements but, after the war, there was a shortage of suitable wood. As a result, the L.M. Ericsson wall telephone was developed

Fig. 8.21 *A 1930s Neophone, to which a bell in a moulded case has been added as a plinth*

Fig. 8.22 *This Neophone, the colour of old gold and specially inscribed, was installed for King George V in 1931 as the two millionth telephone connected to the British Post Office system*

Fig. 8.23 *A Neophone in the map room of the Cabinet War Rooms under London from which, in spite of air raids, Winston Churchill directed the British war effort during the Second World War. The remains of a circular label warns that without a scrambler, speech may be overheard*

in Britain. This development closely paralleled the British conversion of the Scandinavian table telephone. Once again, the case of the instrument was combined with the Neophone handset and the composite telephone became one of the '300' range. It was introduced in 1950 and superseded five years later.

The introduction of the one-piece moulded plastic telephone case revolutionised the appearance of telephones. However, it did not achieve its supremacy without challenge. There were many competing designs, some of which incorporated novel and interesting ideas.

In America, in the late 1920s, the 'French' telephone evolved through a series of developments. One was the conversion of its base from round to elliptical, the long axis of the ellipse being across the width of the instrument. This produced a surface which was more flat at the front where the dial was mounted and enabled the two shapes to be more easily blended into each other. The 'French' telephone was also made in various surface finishes including grey, ivory, statuary bronze, gold and oxidised silver. The end result was an attractive and popular design. The success of this telephone led to a similar instrument being made in Britain by the Automatic Telephone Manufacturing Company of Liverpool. This company was associated with the Automatic Electric Company of Chicago, USA – one of

Fig. 8.24 *1929 table telephone made by Siemens and Halske of Germany using a handset designed by Hasler of Switzerland*

the makers of the telephone in America. The British telephone consisted of a body with a rounded base, similar to the original American design, combined with a Neophone type handset. Today this telephone has acquired the nick-name 'The British version of the American "French" telephone'. This name is not only an apt description of the international nature of this telephone's design, it also symbolises the international nature of telephone design in general. However, at the time this telephone was made it was one of a series of wall and table telephones marketed under the trade name 'Strowgerphone'. It is therefore more correctly called the 'table Strowgerphone'.

In America, where the 'French' telephone was the first handset telephone to be used by the Bell Company, it was an unqualified success. But in Europe, where handset telephones were by no means novel, this success was not repeated. It failed to be adopted by any major telephone administration and was mainly used in private telephone systems installed by the Automatic Telephone Manufacturing Company and also as a railway telephone.

In Germany an extremely compact table telephone was developed. It had a circular base only slightly larger in diameter than the dial. The distinctive shape

Fig. 8.25 *Sound-powered telephone made by the Telephone Manufacturing Co. of Dulwich, London, and used by Allied Forces during the Second World War*

Fig. 8.26 *Telephone designed by Elektrisk Bureau of Norway, developed and manufactured by L.M. Ericsson of Sweden in 1931. This was the first telephone to have a one-piece moulded case*

Fig. 8.27 *The British Post Office '300 series' table telephone brought into general use in 1937. It combined the case designed and developed in Scandinavia, with the Neophone handset. Because its sloping front resembled a dish for cheese, it was nicknamed the 'cheese dish' telephone*

Fig. 8.28 *A '300 type desk set' used by the Bell Telephone Company of America. Originally designed in 1937 by the Western Electric Company, from 1940 onwards it was manufactured with a one-piece moulded plastic case*

of its body earned it the nick-name 'Kuhfub fernsprecher' – 'Cow's hoof telephone'. If the story in the musical show 'Cabaret' were true and there really had been a Blue Angel night club in Berlin, this is the type of telephone that would have been fitted on the tables.

Telephones were also developed in Germany with a novel form of dial. Ten digits were arranged on a backward curving surface in two vertical rows of five,

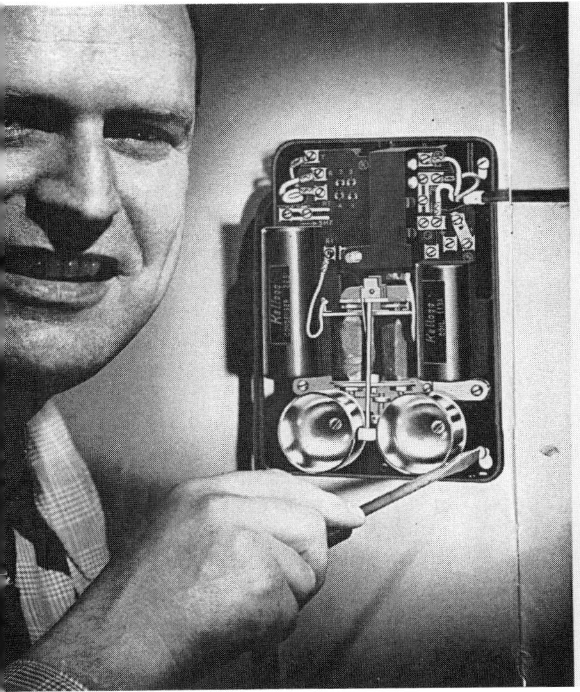

KELLOGG Wasn't Kidding!

. . . they're a cinch to install!

That's what your installer will say as he installs your new Kellogg 1000 Series Masterphones. Just watch him! The only tool he uses—the only tool he needs—is an ordinary screwdriver. No circuit-wiring to solder—no complicated color-codes to follow. No muss or fuss to delay the job.

Routine maintenance is a cinch, too. Plug-in, universal-type condenser and coil; plug-in dial connection; universal, enclosed circuit and easily-adjusted ringer speed up inspection and make adaptations to any service need easier than ever before!

Give your subscribers the finer reception, transmission and ringing—the greater dependability—the eye-appealing, modern styling of Kellogg 1000 Series Masterphones. Give your company the increased subscriber satisfaction, lower maintenance and inventory costs these instruments make possible. Mail your order *today!*

KELLOGG MASTERPHONE

THE TELEPHONE THAT'S YEARS AHEAD

KELLOGG SWITCHBOARD AND SUPPLY COMPANY
6650 SOUTH CICERO AVENUE • CHICAGO 38, ILLINOIS

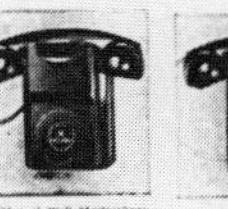

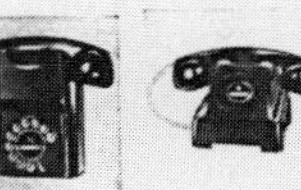

Fig. 8.29

Economize on Maintenance—
Standardize on

MONOPHONES

Use Monophones *exclusively*, and your maintenance costs hit bottom! For two good reasons:

First—Monophones are TOUGH!

We know the kind of handling a telephone gets in service—and we build Monophones to take it! Plastics for housings are especially selected for their resistance to breakage, surface abrasions and loss of brilliance. The Monophone transmitter and receiver are "shockproofed" to withstand the banging of the handset, without injury. The dial is Automatic's famous long-life dial. And so it goes, with every part of these rugged instruments.

Second—Monophone Parts are STANDARDIZED

All regular Monophones use the same handset, the same dial, the same transmitter, receiver, induction coil, ringer and condenser. With these few items in your stockroom, you are ready to make replacements on any of your Monophones.

Install Monophones, and you'll know why so many telephone companies have found that "We can save money by standardizing on Monophones." It's the one sure way to beat maintenance costs!

Standardize on Monophones—starting NOW.

AUTOMATIC ELECTRIC

AUTOMATIC ELECTRIC SALES CORPORATION, 1033 W. Van Buren St., Chicago 7, U. S. A.
INTERNATIONAL AUTOMATIC ELECTRIC CORPORATION

Type 40 Desk Model Monophone—durable, sleek in appearance.

Type 43 Compact Monophone for desks, wall or post mounting.

Type 50 Wall Mounting Monophone, noted for its convenience and decorative appeal.

Type 44 Wall Monophone—sturdy, efficient utility instrument.

Fig. 8.30 *1940s advertisements for telephones with one-piece moulded cases used by the independent companies in the USA*

Fig. 8.32

1940s 'advertisements for telephones with one-piece moulded cases used by independent companies in the USA'

Fig. 8.33 *This wall telephone combined a one-piece moulded case, developed in Scandinavia in 1931, with a Neophone handset. This combination was introduced with new circuitry by the British Post Office in 1950*

Fig. 8.34 *Table Strowgerphone inspired by the American 'French' telephone and made by the Automatic Telephone Manufacturing Company of Liverpool, England, in the 1930s. This telephone was used in Australia*

Fig. 8.35 *An A.T.M. advertisement of 1934 showing (from the top) a Neophone, a wall Strowgerphone, a Neophone with bell and a table Strowgerphone*

with the odd numbers on the left and even numbers on the right. Associated with each number was a depression into which the caller could place his finger. To dial a number, the caller pulled the depression downward until it reached a finger stop and then released it. Although this device was superficially quite unlike a telephone dial, when it was removed from the telephone it was apparent that it was, basically, a conventional dial turned on its edge.

In 1935 a new space-saving telephone was developed in Scandinavia and manufactured by L.M. Ericsson. It was specifically called the 'Domestic' telephone and was intended for use in the home. It consisted of a moulded

Fig. 8.36 *Made in the 1930s, this German telephone was nick-named 'kuhfub fernsprecher' – 'cow's hoof telephone'*

handset hung by its earpiece from an extremely compact wall-mounted bracket which incorporated a gravity switch. As a domestic telephone it was not a success. However, a similar telephone was made in Britain by the Plessey Company at its Ilford, London factory. This instrument was fitted with a Neophone handset and called the 'Pendant' telephone. Unlike the Scandinavian instrument, it was intended for use in the home but was marketed as a business executive's telephone. The advantage claimed for it was that it could be fitted into the knee-hole of a desk, leaving the desk top completely clear for paperwork. In its new capacity it was moderately successful and remained in use in its modified form for several decades. The dial was contained in a separate unit in the form of a drawer attached to the underside of the desk top. To make a call, the drawer was pulled out to reveal the dial; at other times the drawer remained under the desk. This telephone was also one of the first to have an elasticated cord.

The repercussions from the introduction of plastics and the one-piece telephone case, have lasted until the present day. Instruments designed before these developments can justifiably be described as 'vintage' or 'early' types, while those incorporating the one-piece plastic case or later developments are more aptly called 'second generation' telephones.

Early telephones have been tested by time and experience and we can judge them with the benefit of hindsight. Furthermore, we can make these judgements without being influenced by the fashions or trends of the time. It is also possible, in retrospect, to see the overall evolutionary pattern and to recognise which designs contributed to the main line of development and which were merely off-shoots or even retrograde steps.

Fig. 8.37 *The development of this unconventional German telephone was delayed by the war and it did not come into use until peace was restored*

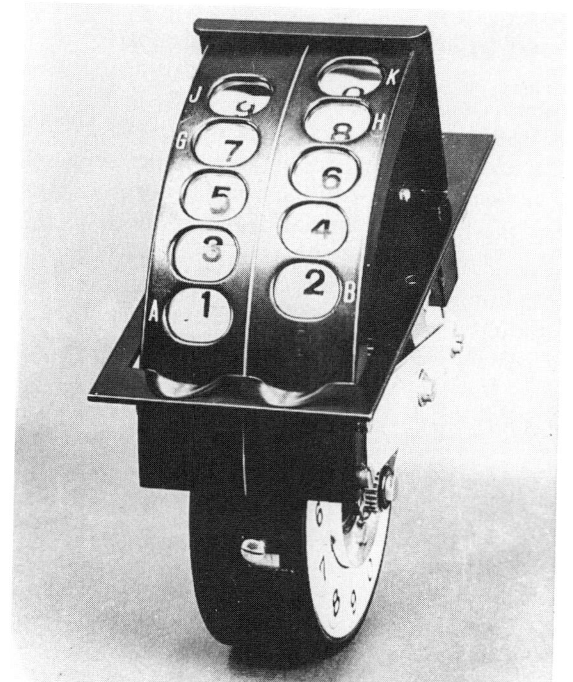

Fig. 8.38 *Unconventional as the dial from the German telephone may seem, when it is removed from its case it is immediately recognisable as an ordinary dial turned on its edge*

Fig. 8.39 *The pendant telephone, conceived in Scandinavia in the early 1930s, was manufactured for the British Post Office by Plessey Ltd. of Ilford, England, in 1938. It was intended for executives who appreciated an uncluttered desk top*

When we assess the merits of second generation designs, including today's telephones, we have no benefit of hindsight. Nevertheless, many of the standards by which early telephones can be judged, hold good. For example, it is still true that a telephone should, first and foremost, be fit for the job it has to do. Under this heading can be included such things as the convenience and comfort of the telephone user. Telephones should also be attractive in appearance and unpretentious. In short, the sort of qualities which are sometimes summed up as 'good taste'.

References

1 L'Illustration, 26 August 1854.
2 Le Téléphone by Count du Moncel.
3 U.S. Specification No. 202,495 dated 16 April 1878.
4 U.S. Specification No. 222,201 dated 2 December 1879.
5 U.S. Patent No. 463,569 dated 17 November 1891.
6 U.S. Patent No. 221,958 dated 25 November 1879 and British Patent No. 5,335 dated 31 December 1879.
7 Preface to 'The Irrational Knot' by George Bernard Shaw, published by Archibald Constable & Co. Ltd., London, 1895.
8 Talk to the Royal Society by Professor Huxley on 8 May 1878.
9 British Patent Specification No. 412 dated 1 February 1879.
10 U.S. Specification No. 186,787 dated 30 January 1877.
11 The Electrical Review, London, X, 350.
12 British Patent Specification No. 3647 dated 16 September 1878.
13 'Bericht über die Electrische Ausstellung', Vienna 1883.
14 Patent Specification No. 5276 dated 15 April 1886.
15 British Patent No. 2451 dated 24 May 1882.
16 British Patent Specification No. 11304 dated 15 July 1889 and Specification No. 21565 dated 9 December 1891.
17 U.S. Patent No. 224,138 filed 29 September 1879 and issued on 3 February 1880.
18 U.S. Patent No. 327,073 granted to H. Edmunds and C.T. Howard, applied for on 13 May 1885 and issued on 29 September 1885.
19 'Deep Diving and Submarine Operations' by Sir Robert H. Davis, 1935.
20 British Patent No. 2488 granted to C.F. Antell, 5 February 1901.
21 British Patent No. 21,574 granted to N.J. Tibbs, R.H. Smith and C.T. Howard, and U.S. Patent No. 327,073 applied for on 13 May 1885 and issued on 29 September 1885.
22 U.S. Patent No. 243,165 dated 21 June 1881 granted to C.E. Scribner, and British patent of 1882 granted to George Lee Anders.
23 British Patent No. 7081 issued to H. F. Hansell on 4 February 1909. (A more rudimentary receiver holder was described in Patent No. 10,559 issued to G.S. Meyer on 8 December 1900.)
24 U.S. Patent No. 222,458 applied for on 10 September 1879 and granted on 9 December 1879, and British Patent Specification No. 5114 dated 13 December 1879.
25 Article by M. Rothen in the 'Journal Télégraphique' of 25 December 1881.
26 U.S. Patents No. 223,201 and 223,202 applied for on 11 & 13 October 1879 and granted on 30 December 1879, and British Patent Specification No. 404 dated 29 January 1880. (Later U.S. patents granted to Westinghouse for automatic switching systems were No. 224,565 of 17 February 1880 and No. 237,222 of 1 February 1881.)
27 British Patent Specification No. 3380 dated 7 July 1883.
28 British Patent Specification No. 7850 dated 6 May 1891, and U.S. Patent No. 447,918 applied for on 12 March 1889 and granted on 10 March 1891.
29 A more detailed explanation of the Strowger system is contained in 'The Telephone and the Exchange' by P.J. Povey, published by Pitman.
30 British Patent No. 308,630.

Picture credits

Belgium
Bell Telephone Manufacturing Company, Antwerp:
Fig. 7.2.

Brazil
Telemig:
Fig. 7.29.

Denmark
KTAS, Copenhagen:
Figs. 6.26, 8.12.

France
Centre National d'Études des Télécommunications:
Fig. 4.4.

Historical Collection of French Telecommunications:
Figs. 2.6, 2.26, 3.4, 6.10, 6.12, 6.13, 6.17, 7.19, 7.21, 7.22.

German Federal Republic
Bundespostmuseum, Frankfurt:
Figs. 1.8, 1.9, 3.23, 3.25, 6.9, 6.14, 6.15, 6.16, 6.18, 6.27, 6.35, 8.36, 8.37, 8.38.

Siemens Museum, Munich:
Figs. 3.24, 3.26

Great Britain
E. T. Birch, Liverpool:
Fig. 5.9.

British Telecom:
Figs. 1.2, 1.5, 1.12, 1.16, 2.1, 2.5, 2.8, 2.19, 2.27, 3.21, 3.22, 3.29, 4.5, 4.6, 4.8, 4.9, 4.11, 5.5, 5.12, 5.13, 5.14, 6.2, 6.6, 6.8(2), 6.21, 6.23, 6.32, 6.38, 6.48, 6.51, 7.10, 7.27, 8.4, 8.9, 8.10, 8.11, 8.13, 8.14, 8.15, 8.18, 8.19, 8.20, 8.21, 8.22, 8.27, 8.39.

Great Britain (Cont'd)

R.A.J. Earl, Oxford:
Figs. 1.4, 2.3, 2.7, 2.25, 3.5, 3.30, 3.32, 4.7, 5.23, 5.24, 5.25, 5.26, 5.28, 5.29, 6.7, 7.4, 7.5, 7.6, 7.7, 7.8, 7.16, 7.17, 7.20, 7.23, 7.24, 7.25, 7.26, 7.28, 7.36, 7.38, 7.39, 8.1, 8.2, 8.16, 8.25, 8.33, 8.34.

Imperial War Museum, London:
Fig. 8.23.

By kind permission of the Marconi Company Ltd.:
Fig. 4.12.

Mary Evans Picture Library:
Figs. 4.3, 4.10, 5.1, 6.1.

Plessey Major Systems Ltd.:
Figs. 7.30, 7.31, 8.32.

P.J. Povey, Isle of Wight:
Fig. 5.10.

Science Museum, London:
Fig. 1.10.

Netherlands
Netherlands Postmuseum, The Hague:
Fig. 6.11.

Phillips TMC:
Figs. 7.13, 7.14, 7.15.

New Zealand
Telecom Corporation of New Zealand Ltd.:
Fig. 1.11.

Norway
Norske Elektrisk Bureau:
Figs. 3.33, 8.26.

Portugal
Museu dos Correios e Telecomunicações:
Figs. 2.4, 6.5.

Sweden
Ericsson, Stockholm:
Figs. 1.6, 1.7, 2.16, 2.17, 2.21, 2.22, 2.35, 3.1, 3.2, 3.3, 3.16, 3.20, 6.3.

Switzerland
PTT Museum:
Fig. 8.24.

USA
AT & T:
Figs. 1.13, 2.23, 3.18, 3.19, 8.17, 8.28.

Reprinted by permission of Bell Telephone Laboratories Incorporated, copyright 1975:
Fig. 3.15.

Willard R. Culver, copyright 1947, National Geographic Society:
Fig. 1.15.

GTE Corporation:
Figs. 6.37, 6.46, 6.47, 6.49, 8.30.

ITT Corporation:
Fig. 8.29.

National Museum of American History, Washington:
Fig. 3.28.

Northern Telecom:
Figs. 7.18, 8.31.

Northwestern Bell:
Figs. 6.19, 6.42, 6.45.

Index